STAINING OF
FACADES

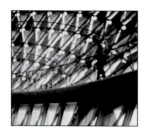

STAINING OF
FACADES

Michael Y L Chew
Tan Phay Ping

School of Design & Environment, NUS

 **World Scientific**

NEW JERSEY · LONDON · SINGAPORE · SHANGHAI · HONG KONG · TAIPEI · BANGALORE

Published by

World Scientific Publishing Co. Pte. Ltd.

5 Toh Tuck Link, Singapore 596224

USA office: Suite 202, 1060 Main Street, River Edge, NJ 07661

UK office: 57 Shelton Street, Covent Garden, London WC2H 9HE

British Library Cataloguing-in-Publication Data
A catalogue record for this book is available from the British Library.

STAINING OF FACADES

ISBN 981-238-298-4
ISBN 981-238-299-2 (pbk)

Printed in Singapore by Mainland Press

PREFACE

Millions of dollars are spent annually to remove stains from the facades of buildings. To shed the image of construction industry as being "backward and low in productivity", buildings would have to be designed and built to be more maintainable so that resources spent on maintenance works could be minimised.

The facade is the building skin that encloses and protects the occupants from weather elements, and it also gives the first impression for the building. With increasing awareness on minimising life-cycle costs, building clients and owners are specifying for buildings that require minimal maintenance costs. Facades that are designed without consideration for locational influences, dirt retention abilities, rainwater runoff drainage and access for cleaning and maintenance will often experience premature staining and incur high cleaning and maintenance costs. An in-depth understanding on the factors that degrade the visual image of the building because of facade staining is therefore important.

This book aims to highlight the various factors that need to be considered during the design, construction and post occupancy stages to minimise the problems with staining. Facade elements studied include exposed brickwork, concrete, natural stones, tiles, metal, glazing and plaster/painted walls. The implications of environmental, material, design and maintenance aspects with regards to staining of facades are discussed.

Chew Yit Lin, Michael
Tan Phay Ping
National University of Singapore

ACKNOWLEDGEMENT

We would like to express our gratitude to the following people for their assistance and support.

Building and Construction Authority, Singapore

Research team members:
- Tan Shan Shan
- Nayanthara De Silva
- Seng Soon Huat, Bernard
- Wang Tung Shu
- Petikirige Ranga Yashoman Edir
- Ng Chew Peng, Camelia
- Chan Huan Teng

Industry partners:
- Ms Wee-Lim Eng Hon – Diethelm Industries Pte Ltd
- Mr Jean Edouard Perrine (Eddy) – Permatsteelisa Pacific Holdings Ltd
- Mr Michael Lim – Centre for Cleaning Technology Pte Ltd
- Mr Sunny Teo – Tractel Singapore Pte Ltd
- Mr Stefan Gisin – Alcan Alucobond (Far East) Pte Ltd
- Mr David Lim – Integral Components Equipment Pte Ltd
- Mr Wong Chung Wan – Setsco Services Pte Ltd
- P. P. Choo and Goh Joo Wee – MHE-Dematic (S) Pte Ltd
- Eddie Loh and Cedric Yip – Campaign Cleaning Services (Pte) Ltd
- Steven Tan – Shaw Organisation
- Tan Teck Bin – JTC Corporation
- Dr D. Martin-Vosshage – Fraunhofer Representative Office Singapore
- Mr Kang Kok Hin – Individual Capacity

CONTENTS

CHAPTER 1

INTRODUCTION

Staining or premature staining on building facades of new buildings has gained much attention in recent years. Resources required to tackle the problem represent a significant portion of the maintenance expenditure in the local construction industry.

Staining is defined as a mark or discolouration that is not easily removed. The build-up of stains on facades at a time before it is normal or expected can be termed as premature staining [1]. Factors that contribute to staining include material, exposure, design, colour, water absorptivity, dirt retention, texture and solubility. Stains form primarily from the surface flow of water (*runoff*) down the facade. Water brings along dirt particles that is retained on the facade material. When the water dries out, dirt particles that are not washed off manifest as stains [2–12]. Although traditional facade materials such as bricks, stones, concrete and wood are still in use, many non-porous and non-adsorptive materials such as glass, plastic and metal are gaining popularity. They are able to offer greater flexibility and functionality as well as better performances in aspects such as visual, thermal, spatial, acoustic, indoor air quality and building integrity. With new materials and more sophisticated facade designs, the provision of joints to effectively control rain runoff flow on the facade has to be re-looked since these materials are impermeable and runoff is generated almost immediately after a rainfall. New facade designs that give more consideration to runoff flow needs to be advocated. Buildings need to be designed to be more maintainable so that resources used on cleaning and maintenance work may be minimised [13–16]. The inclusion of certain architectural design features may cause retention of dirt and the flow of runoff to be uneven and uncontrolled, resulting in stains.

This book discusses the problems associated with staining from the following aspects:

- Environmental
- Material
- Design
- Maintenance

In Chapter 2, the significance of environmental conditions which a building facade is exposed to in relation to staining is discussed. Environmental conditions discussed include the effect of rain, wind, sunlight and pollutants.

Chapter 3 discusses the susceptibility of a material to staining from the perspective of the material characteristics including permeability, water absorption, surface texture, colour, chemical and biological resistance.

The design aspects are discussed in Chapter 4. The rain runoff flow pattern over some main facade design features is illustrated. The relationship between runoff flow pattern and staining patterns of four main design features namely, ledges, joints misalignment, protruding fixtures and louver units is elaborated. The importance of a combined effort from the design team to use stain-free designs, the construction team to ensure good workmanship, and the maintenance team to implement optimum maintenance strategy is highlighted.

Chapter 5 discusses the whole life performance of a facade in relation to maintenance aspects. Maintainability issues including cleaning, repair and replacement are evaluated. The various facade access systems for tall buildings are illustrated.

References

[1] P. Parnham, Prevention of Premature Staining of New Buildings, *E. & F.N. Spon,* London, 1997.
[2] L. Addleson and C. Rice, *Performance of Materials of Buildings,* Butterworth-Heinemann, Oxford, 1994.

[3] O. Beijer, *Weathering on External Walls of Concrete,* Swedish Concrete Research Council, Swedish Cement and Concrete Research Institute, Stockholm, 1980

[4] C. Briffett, The Performance of External Wall System in Tropical Climates, *Energy and Buildings*, Netherlands, 1990.

[5] C. Briffett, "External finishes — Case studies on problems and solutions", Building Protection Conference, Proceedings of *Inter-Faculty Conference on Protection of Buildings from the External Environment,* Paper 2, 1987

[6] C. Hall, "Absorption and shedding of rain by building surfaces", *Building & Environment* Vol. 17, 257–262, 1982.

[7] R. Cooper, "Factors affecting the production of surface runoff from wind-driven rain. *RILEM International Symposium*, Rotterdam del. 1.1.2, 1974.

[8] L. G. W. Verhoef, *Soiling and Cleaning of Building Facades*, Report of the Technical Committee 62 SCF, RILEM, Chapman and Hall, London, 1988.

[9] M. B. Ullah, "Analysis of rain and wind for building design", *National University of Singapore Seminar on Wind & Rain Penetration in Buildings,* 1994.

[10] M. C. Baker, "Rain deposit, water migration and dirt markings on buildings", *RILEM/ ASTM/ CIB Symposium on Evaluation of the Performance of External Vertical Surfaces of Buildings*, pp. 57–66, 1977.

[11] H. P. Teo and P. Ng, "External cladding defects in Singapore" in *Southeast Asia Building,* March, 1992.

[12] W. H. Ransom, *Building Failures: Diagnostics and Avoidance*, E. & F.N. Spon, London, 1987.

[13] M. Y. L. Chew, C. W. Wong and L. H. Kang, *Building Facades: A Guide to Common Defects in Tropical Climates*, World Scientific Publishing, Singapore, 1999.

[14] M. Y. L. Chew, "Efficient maintenance: Overcoming building defects and ensuring durability," *Conference on Building Safety*, The Asia Business Forum, Kuala Lumpur, 4 & 5 April 1994.

[15] E. B. Feldman, *Building Design for Maintainability*, McGraw-Hill, New York, 1975.

[16] N. G. Marsh, "The effect of design on maintenance", *Development in Building Maintenance — I*, Applied Science Publications Ltd., New York, 1979.

CHAPTER 2

ENVIRONMENTAL ASPECTS

2.1 General

External environmental conditions have a profound effect on the type and extent of staining on a building facade. This is especially true for deterioration processes which are initiated by rain runoff. Without rain runoff, the staining agents (dirt) would probably affect the face of the building evenly [1]. However, the rate, volume and flow pattern of rain runoff flowing on a facade is determined largely by the synergistic workings of rain, wind and sunlight. In averting staining problems, it may be useful to understand the manner by which pollutants in the atmosphere are transported to the buildings and adhered to the facade surfaces, and the ways in which dirt is subsequently redistributed by rain runoff over the facade. Therefore, before the design team embarks on the design, materials and system selection for the facade, they should consider carefully the atmospheric conditions the building is going to be exposed to including:

- Rain
- Wind
- Sunlight
- Pollutants

With a good understanding of the environment (its microclimate and its pollution) in which the building is to be built, a risk analysis can be made right from the planning stage, and this enables the design team to identify areas that require special attention [2]. Materials that are suitable for these situations can then be selected and design features that are able to control runoff flow be incorporated.

2.2 Rain

Rainwater plays a central role in many of the destructive mechanisms that a building experiences, such as dimensional change, corrosion, leaching, efflorescence, and water penetration, which may lead to the deterioration of internal finishes [3]. In the staining of facades, rainwater is the initiator of the process. Movement of water over the facade of the building will result in either washing or deposition of dirt, causing the disfigurement of the building's appearance if such effects are concentrated at different locations. Water is also a major factor in leaching and efflorescence processes, as well as in promoting biological growth, all of which will result in the staining of the building facade.

From a study carried out on the rainfall data for a period of 10 years (1988–1997), it was observed that Singapore received an average of 238.5cm of rainfall [4]. Figure 2.1 shows the 5 major regions of Singapore (North, South, East, West and Central) and Figure 2.2 shows the average rainfall in these respective regions. It appears that there is little difference in the average rainfall level between each region for the period 1988–1997, with the lowest rainfall recorded in the East.

Hence, in Singapore the general location of a building does not affect the amount of rainfall and the extent of the washing effect that it will receive.

2.3 Wind

Most rainfalls are accompanied by wind, which makes it important to consider a building's exposure to the prevailing wind direction. One fundamental effect of wind during rainfall is that it will change the direction of fall of raindrops so that they impinge on vertical surfaces instead of falling parallel to the surfaces [5,6,7]. Wind force does not only bring about driving rain but may also alter the pattern of runoff flow

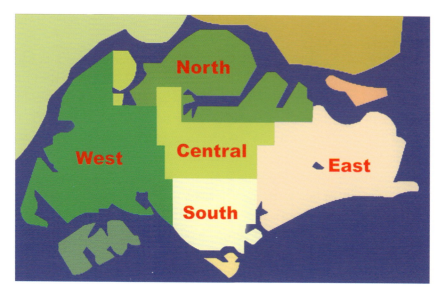

Figure 2.1. Map of Singapore divided into 5 major parts [4].
(Source: http://intranet.mssinet.gov.sg/nowcast/)

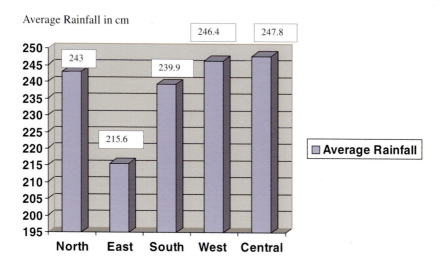

Figure 2.2. Average rainfall in the 5 parts of Singapore [4].

on the facade. Since the direction from which the wind blows changes throughout the year, the direction from which rain will impinge on the building facade will vary accordingly.

From a study conducted to determine the predominant direction and the speed of wind in Singapore, it was found that over a period of 10 years (1988–1997), the general prevailing wind direction for all sectors is in a N to NE sector direction, followed by the S to SE sector direction (Fig. 2.3). The range for the 10-year average wind speed is small, ranging between 1.6 to 2.3 m/s. Wind from the N and NW sectors have generally lower wind speed (Fig. 2.4).

The quantity of rainwater, the velocity and the angle at which rain hits the building varies significantly at different parts of a building. Wind driven rain is defined as "rain carried along by wind at an angle to the vertical" [5,6,7]. In Singapore, rainfalls are usually accompanied by wind with speeds of between 1.6 to 2.3 m/s. Generally, winds from N and NW sectors have lower wind speeds (Fig. 2.4). This wind speed increases significantly with respect to the height of the building. Wind speeds of up to 20 m/s have been recorded on the 20th storey of buildings in various parts of Singapore [6].

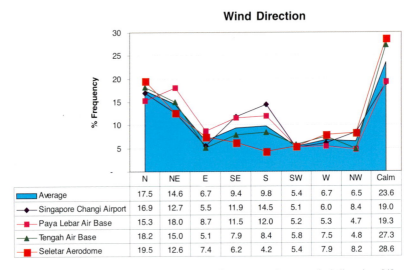

Wind Direction

	N	NE	E	SE	S	SW	W	NW	Calm
Average	17.5	14.6	6.7	9.4	9.8	5.4	6.7	6.5	23.6
Singapore Changi Airport	16.9	12.7	5.5	11.9	14.5	5.1	6.0	8.4	19.0
Paya Lebar Air Base	15.3	18.0	8.7	11.5	12.0	5.2	5.3	4.7	19.3
Tengah Air Base	18.2	15.0	5.1	7.9	8.4	5.8	7.5	4.8	27.3
Seletar Aerodome	19.5	12.6	7.4	6.2	4.2	5.4	7.9	8.2	28.6

Figure 2.3. 10-year averaged percentage frequency of mean wind direction [4].

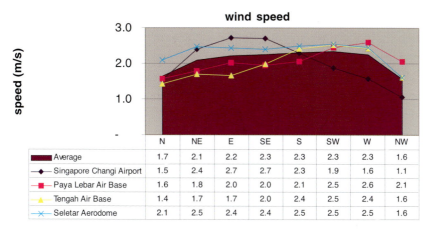

		N	NE	E	SE	S	SW	W	NW
▬	Average	1.7	2.1	2.2	2.3	2.3	2.3	2.3	1.6
◆	Singapore Changi Airport	1.5	2.4	2.7	2.7	2.3	1.9	1.6	1.1
■	Paya Lebar Air Base	1.6	1.8	2.0	2.0	2.1	2.5	2.6	2.1
▲	Tengah Air Base	1.4	1.7	1.7	2.0	2.4	2.5	2.4	1.6
✕	Seletar Aerodome	2.1	2.5	2.4	2.4	2.5	2.5	2.5	1.6

Figure 2.4. 10-year averaged surface wind speed [4].

The dominant wind direction affects how rain will strike the facade and how the resultant rain runoff will flow and redistribute the dirt on the facade. Beijer (1980) [7] reported that:

- Less than half the quantity of rain that should pass through an equivalent cross-sectional area of "free air" is caught by the external wall. This applies regardless of the wind force.
- The top parts of the external wall receive much more rain than the lower parts.
- Raindrops move almost parallel to the lower sections of the external wall.

From such analysis it is expected that in Singapore, wind-driven rain is predominant from N and NE directions. Facades orientated towards these directions may receive more rain impacting on them, resulting in greater washing and deposition of dirt. Furthermore, the angle made between the raindrop and the vertical surface will be greater for facades facing SE, S, SW and W due to the greater speed of wind at these orientations. This may give rise to differing staining patterns due to sheltering effect especially on facades that have projections.

When wind moves across a building, it changes its direction especially around the edges of a building. Raindrops that are driven by the wind would not be able to follow this directional change and would strike the facade. Hence, the wetting pattern for a tall building would be heavier at windward corners, projections and parapets. The washing effect at such areas would thus be relatively stronger [7].

Once the raindrops impinge on the facade and generate enough runoff to run over the face of the wall, its flow would be rather vertical but this is also subjected to the force of lateral winds that has the ability to alter the pattern of flow [3]. This wind force increases as the building height increases.

2.4 Sunlight

Sunlight plays an important role in affecting the formation of stains on a facade. Some surfaces may be exposed directly to long periods of sunshine while others may be sheltered and remain in constantly damp conditions. The intensity and duration of sunlight that a surface receives affect the limit of runoff flow, the type of biological stains and hence the pattern of staining. A facade's exposure to sunlight may be affected by its orientation since the north and south facing facade generally receive less sunlight compared to the east and west facing ones [8,9,10,11]. The presence of adjacent buildings and other structures may also provide a sheltering effect. These may result in the facade having a slower drying process after it has been wetted, leaving it damp for longer periods and therefore promoting biological staining. The orientation and location of facades thus may have implications on the extent of its staining.

Adopting the horizontal coordinate system, the direction that the sunlight strikes the earth can be represented by the solar altitude angle and the solar azimuth angle. The solar altitude angle describes how high the sun appears in the sky (Fig. 2.5), and is measured between an imaginary line drawn between the observer and the sun and the horizontal plane the observer is standing on. This angle is negative when the sun drops below the horizon. The solar azimuth angle is the angular distance between due South and the projection of the line of sight to the sun on the ground (Fig. 2.6). A positive solar azimuth angle indicates a

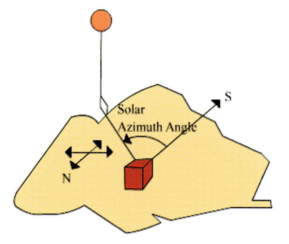

Figure 2.5. Solar altitude angle.

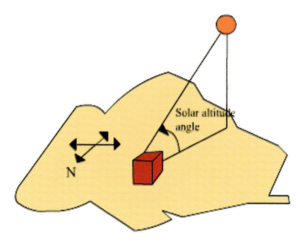

Figure 2.6. Solar azimuth angle.

position East of South, and a negative azimuth angle indicates a position West of South. The angles vary at different times of the day and at different locations due to the earth's orbit around the sun and its own orbit on its axis [8].

Solar altitude angles and azimuth angles have been measured extensively in various independent studies carried out to determine sky luminance in Singapore [8,9,10,11]. In one such study on sky luminance of Singapore in 2000, it was found that solar altitude angle increased from 0700 hrs and peaked at about 1300 hrs in the afternoon with a value of 75 degrees (average) before decreasing as it approached 1900 hrs [11]. The solar azimuth angle on the other hand was found to be about 90 degrees (average) at 0700 hr indicating that the sunlight was shining from the east and was about −90 degree (average) at 1900 hrs indicating that the sunlight was shining from the west. Generally, for the months of April to September 2000, facades facing North received sunlight mainly from ENE and WNW directions and from the North directly for a short period in the afternoon while facades facing the South did not receive any sunlight. For the months of October to March 2000, facades facing South received sunlight mainly from ESE and WSW directions and from the South directly for a short period in the afternoon while facades facing the North did not receive any sunlight. The findings were supported by results from actual measurements [9, 10]. North and South facing surfaces remain wetter for longer periods after wetting as compared to the East and West facing facades.

In addition, for every part of a facade to receive sunlight daily (assuming a yearly minimum solar altitude angle of 65 degrees [11], the adjacent building's height must also not be more than 2.14 m high for every 1 m the adjacent building is away from the facade (Fig. 2.7).

2.5 Pollution

Singapore uses the guidelines of World Health Organization (WHO) and the United States Environmental Protection Agency (USEPA) in assessing its pollution levels. Since 1990, the concentration of pollutants in the air were generally at very low levels and the Pollutant Standard Index (PSI) fluctuated between good and moderate levels [12]. The complex process of staining begins with the deposition of staining agents on the facade. There are two types of staining agents that can affect

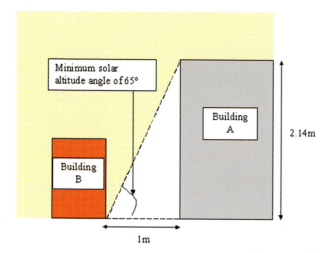

Figure 2.7. Permitted height of adjacent building if a facade is to receive sunlight daily.

building facades: non-biological and biological staining agents. These staining agents may be carried by the wind and deposited directly on the facade or they can be deposited in solution by the rainwater.

2.5.1 Non Biological Staining Agents

Non-biological staining agents that build up as dirt stains on buildings are a result of air pollution from human activities. Pollution sources that emit pollutants into Singapore's atmosphere include:

- Industrial plants and presence of pollutants emitting processes
- Vehicular traffic on heavily utilised roads
- Airborne dust from neighbouring countries

Pollutants from these sources can vary in sizes from molecular to particulate. They can be grouped according to their sizes. Table 2.1 shows some common pollutants in air [13,14].

Molecular pollutants such as sulphur dioxide are present in the air and they usually dissolve in rainwater to form acidic rain that corrodes and attacks facade materials causing localised losses of material.

Table 2.1. Common pollutants in air [13, 14].

Pollutants	Examples	Sizes	Sources	Possible Effects
Aerosols	Sulphur dioxide (SO_2).	Gaseous	Burning of fuels.	Formation of acid rain. Reaction with water and other materials to form sulphates and chlorides, causing destruction to facade materials.
Soot or black smoke	Smoke from tobacco, coal, fuel oil, metallurgical industry.	$< 1\mu m$	Imperfect combustion.	Transported by air and deposited on facades, causing stains.
Coarse particulate matter	Ash, dust, rock debris, mineral dust.	$> 1\mu m$	Unburnt fuel and dust from roads and industries.	Transported by air for short distances and deposited on horizontal or sloping surfaces to form stains.

Particulate matters in the air generally comprise dust, smoke and suspended particles. Dust particles being large and heavy, settle quickly and will not affect large areas. Smoke and suspended particles being smaller and lighter remain in the atmosphere for a longer time and affect larger areas. The main sources of particulate matter in Singapore are fuel-burning equipment such as boilers and furnaces, motor vehicles and construction sites. In the assessment of the effects that air pollutants may have on a building facade, it may be important for the design team to consider:

- The conditions of adjacent buildings as it would give an accurate indication of the level of pollutants and their impact on the facades.
- Proximity of proposed building to industrial plants and processes.
- Proximity to main roads and expressways.
- Up-to-date published information on pollution levels.
- Seasonal increase in pollutant concentration due to migrating pollution from neighbouring countries.

2.5.2 Biological Staining Agents

Biological staining agents might also bring about staining to the facade (Fig. 2.8). In Singapore, staining of facades due to biological staining agents can be traced to a number of diverse plant groups such as algae, fungi, mosses, ferns and figs. In land-scarce countries like Singapore where homes and offices are mostly high-rise, the disfigurement of facades of buildings by algae has attracted the most attention [15]. The main reason for this is the disfigurement of the facade and the costs involved in removing such stains.

Algae can be distinguished from fungi by their characteristic green colour due to the presence of chlorophyll. They may also be orange or blue green in colour depending on the amount of other pigments within it. On the other hand, fungi is colourless but may become blackish or coloured as they become reproductive [15,16,17]. Algae thrive best where there is sun, moisture and nutrients. Singapore's tropical climate offers an abundance of sunlight and rainfall for algae to grow. Dirt that is blown off the road and retained on the facade becomes a source of nutrients. As such, algae are found mainly outdoors on external facades. Species of algae that may be found on the facades of Singapore's buildings are *trentepohlia odorata, Chlorococcum* and an *Alga* comprising of *Scytonema, Schizothrix, Anacystics* (Table 2.2) [18,19].

On surfaces shaded from the direct rays of the sun and where humidity level is high, fungi rather than algae are the organisms that are likely to proliferate. Unlike algae, fungi are unable to synthesise their own organic food and thus have to rely on the medium on which they are growing to provide the nutrients necessary for their growth. Also, fungi begin to grow when the relative humidity of the surface exceeds 70%. For these reasons, they are usually found on internal surfaces of buildings such as bathrooms, kitchen, ceiling boards, wallpapers and glass panels [20].

Other than algae and fungi, biological growth on building facades may also be in the form of mosses, ferns or small plants (Table 2.2). Mosses are small plants that form mat-like patches on walls of old buildings [16]. Ferns and other small plants growing on walls are much

Table 2.2. Common types of biological agents found commonly on external walls of buildings in Singapore [15,16,17,18].

Biological Agent		Characteristics	Conditions for Growth
Algae		When conditions are ideal, algae will appear in 1 to 2 years time. It may appear initially as either light green, blue green or orange coloured filaments or powder which may be slimy when wet. Over time, dirt may collect over the mycelium to result in blackish and conspicuous stains. Common species of algae in Singapore are *Trentepohlia odorata*, *Chlorococcum* and alga comprising of *Scytonema*, *Schizothrix*, *Anacystics*.	Direct and prolonged exposure to sunlight. Surface should be wet during rain. The degree of exposure and the amount of moisture available will dictate the type of algae that will colonise. Sloping surfaces such as ledges of window sills, balconies and verandahs are areas most prone to algae attack. Usually brick, concrete or rendered surfaces that are highly absorbent and textured will support such growth.

Table 2.2. (*Continued*)

Biological Agent		Characteristics	Conditions for Growth
Moss		Patches of small, mat-forming plants. It traps dirt and soil over time to result in a black carpet. Has good water holding ability due to its thickness.	Humid areas such as cracks that retain moisture and dirt. Usually appear after algae have proliferated. Further improves conditions that are suitable for higher plants such as ferns and small plants.
Ferns		Higher plant species. Roots are easily visible. They retain water well and improve conditions for further biological infestation. Common fern species are *Pteris vittaata*, *Nephrolepis biserrata*.	Usually appear on walls of old buildings. Cracks in walls that have accumulations of soil and abundant moisture provide conducive environments for growth to initiate.

Table 2.2. (*Continued*)

Biological Agent		Characteristics	Conditions for Growth
Other plants		They may be typical ornamental plants usually kept in pots.

They should be removed completely when growth is detected. That area of the wall should be restored by patching and painting to inhibit future growth. | May be found together with moss and ferns.

Growth is hastened by deposition of seeds from nearby potted plants placed along window sills and balconies of the same building. |
| **Strangling/ climbing figs** | | Rapid growth. Roots are usually large in girth and fills up the crack that it originates from. Branch roots usually extend downwards to the ground so as to draw water and nutrients from soil.

Common species of strangling figs include *Ficus religiosa, Ficus microcarpa* and *Ficus benjamina*. Common specie of climbing figs is *Ficus pumila*. | Growth originates from cracks in walls and from ledges. Lack of maintenance of facade will allow figs to thrive and cause serious damage to the facade.

Over time, figs and extending roots may totally cover the building facade. |

more visible forms of biological staining and should be removed prior to further weathering of surface.

Strangling figs are plants that propagate through dispersion by birds. They usually grow out of crevices in walls and their roots extend downwards toward the ground [15]. They should be removed as soon as possible because growth of their roots over time may cause pressure to be exerted on the cracks, resulting in widening of cracks. Climbing figs on the other hand, are usually grown on walls deliberately for aesthetics purposes. They require regular pruning in order to maintain their aesthetic qualities (Table 2.2).

Besides being aesthetically unpleasant, biological growth on facades may also cause deterioration and further weathering to the wall (Fig. 2.9) and if occurring internally, may also manifest to become an environmental health hazard. Preventive maintenance should be carried out to eradicate biological staining. This may involve the use of paint specified to SS 345: 1990 or similar finishes that are designed to prevent or combat biological growth [21,22]. In addition, wall design should include provisions for proper channeling of runoff so that runoff is not concentrated at particular areas.

References

[1] P. Parnham, *Prevention of Premature Staining of New Buildings*, E. & F.N. Spon, London, 1997.

[2] L. G. W. Verhoef, *Soiling and Cleaning of Building Facades*, Report of the Technical Committee 62 SCF, RILEM, Chapman and Hall, London, 1988.

[3] G. Robinson and M. C. Baker, "Wind-driven rain and buildings", Technical Paper 445, National Research Council of Canada, Division of Building Research, Ottawa, 1975.

[4] *Summary of Observations (1988–1997)*, Singapore Meteorological Service.

[5] L. Addleson and C. Rice, *Performance of Materials of Buildings*, Butterworth-Heinemann, Oxford, 1994.

[6] E. C. C. Choi, "Wind-driven rain characteristics and criteria for water penetration test", in Proceedings [of the] *International Conference on*

Building Envelope System and Technology. Centre for Continuing Education, Nanyang Technological Education, 1994.

[7] O. Beijer, "Weathering on external walls of concrete", Swedish Concrete Research Council, Swedish Cement and Concrete Research Institute, Stockholm, 1980.

[8] K. P. Lam, "Mapping of the sky luminance distribution and computational prediction of daylighting performance in Singapore", Research Report, School of Building and Estate Management, National University of Singapore, 1997.

[9] M. B. Ullah, "Energy efficiency performance studies of variable air volume air-conditioning systems under partial load conditions", Staff Research, School of Building and Estate Management, National University of Singapore, 1996.

[10] M. B. Ullah, K. P. Lam, K. W. Tham and P. R. Tregenza, "Study of daylight attenuation through windows in urban environments", Staff Research, Department of Building, National University of Singapore, 2001.

[11] T. S. Wang, and C. W. Toh, "Environmental effect on facades staining in Singapore", Unpublished Student Report, Department of Building, National University of Singapore, 2001.

[12] *Pollution Control Report (1990–1998),* Pollution Control Department, Ministry of the Environment, Singapore.

[13] W. M. Marsh and J. Groosa, *Environmental Geograph: Science, Land Use and Earth System,* John Wiley & Sons, New York, 2002.

[14] C. W. Spicer, *Hazardous Air Pollutants Handbook: Measurements, Properties and Fate in Ambient Air,* Boca Raton, FL: CRC Press/ Lewis Publishers, 2002.

[15] G. Lim, T. K. Tan and A. Tan, "The fungal problems in buildings in the tropics." *International Biodeterioration,* Vol. 25, pp. 27–37, 1989.

[16] N. H. Chua, S. W. Kwok, K. K. Tan, S. P. Teo and H. A. Wong, "Growths on concrete and other similar surfaces in Singapore." *Journal of the Singapore Institute of Architects,* Vol. 51, pp. 13–15, 1972.

[17] Y. C. Wee and R. Corlett, *The City and the Forest: Plant Life in Urban Singapore.* Singapore University Press, Singapore 1986.

[18] Y. C. Wee and K. B. Lee, "Proliferation of algae on surfaces of buildings in Singapore." *International Biodeterioration Bulletin,* Vol. 16, pp. 113–117, 1980.

[19] K. K. Ho, K. H. Tan and Y. C. Wee, "Growth conditions of Trentepholia odorata (Chlorophyta, Ulotrichales)." *Phycologia*, Vol. 22, pp. 303–308, 1983.

[20] A. F. Bravery, "Origin and nature of mould fungi in buildings." In *Proc. Seminar on mould growth in buildings*. Building Research Establishment, Princes Risborough Laboratory, UK, 1980.

[21] P. Whiteley, "The occurrence and prevention of mould and algal growths on paint films." *Society of Chemical Industry Monograph*, Vol. 23, pp. 161–169, 1996.

[22] J. F. Ferguson, "Microbiology of paint films I." *Paint Technol.*, Vol. 33, No. 6, pp. 19–27.

CHAPTER 3

MATERIAL ASPECTS

3.1 General

Material properties such as permeability, water absorption, surface texture, colour, resistance to chemical attacks and resistance to biological growth determine the susceptibility of a material to staining [1]. The understanding of material properties with regard to staining is essential so that a suitable facade material can be selected to withstand the environmental conditions of a particular location.

3.1.1. Permeability

The rates at which liquids and gases can move in a material are determined by its permeability. Permeability affects the way in which a material can resist external attacks and the extent of which the facade structure can be free from leaks. The permeability is much affected by the nature of the pores, both their size and the extent to which they are interconnected [2].

3.1.2. Water Absorption

Water absorption may be explained as the material's ability to absorb and retain water. This property can be quantified by a coefficient known as the absorption coefficient. When the absorption coefficient of the material is lower than a certain value, rain runoff will be formed. The denser the material, the easier and quicker runoff will be generated. For materials such as brick and concrete, runoff rarely reaches the ground with an average rainfall. In the case of metal and glazed surfaces whose absorption coefficient is almost zero, runoff can reach the ground with even weak rainfalls [3, 4].

A facade material with high water absorption is likely to remain damp for a longer time and thus may allow dirt particles to adhere to its surfaces more easily. Porosity and water absorption are closely related. Porosity determines the quantity of liquid or gas that can be contained in the material. Driving rain strikes mostly the top part of a building facade. For facades that are composed of more than one material, it may be important to consider the absorption of the material that makes up the upper portion of the external building wall since it will dictate the amount of runoff that will flow down the building.

3.1.3. Surface Texture

The surface texture of a material dictates the ease of dirt retention on the material's surface, the flow pattern of the runoff and the conspicuousness of the stains. Textured surfaces tend to have higher retention of dirt than a smoother surface. A textured surface may thus experience greater staining due to a larger amount of retained dirt. On the other hand, textured surfaces are better able to mask staining than smooth surfaces. The occurrence of stains would be minimised if water flows evenly over the facade. Textured surfaces have the tendency to break up and disperse the water flow over the surface. The occurrence of stains thus depends on the "streams" created when runoff flows over textured surfaces. In this chapter, surface texture as perceived by the naked eye and not on a microscopic level will be discussed.

3.1.4. Colour

Pigments or admixtures can be added to most materials to achieve the desired colour results. A light-coloured surface shows stain streaks easily and thus it may be wise to use darker shades for areas where stain streaks are likely to manifest (Figure 3.1). Achieving a balance of light and dark

Figure 3.1. Light coloured facades stain easily.

colours throughout a facade can not only enhance the appearance by breaking the monotony of the facade, but also be efficient in masking stained areas.

3.1.5. Resistance to Chemical Attacks

The main pollutants of concern are sulphur dioxide and carbon dioxide which can dissolve in rainwater to produce sulphurous acid and carbonic acid respectively. Both acids may react with salts in building materials.

Besides attacks from environmental pollutants, facade materials may deteriorate due to attacks or adverse reactions from adjacent materials. An example is sealant staining along the joints of natural stone facades. A good knowledge of the reactivity of facade materials and components in the selection process is thus important.

A material's resistance to chemical attack depends very largely on its quality of manufacture and the properties of its constituent materials. The most common form of chemical attack is that due to the sulphates of calcium, magnesium, sodium and potassium which occur widely in clays.

3.1.6. Resistance to Biological Growth

Moisture and sunlight is essential for most forms of biological growth. The extent to which a material may be affected by biological growth is hence determined largely by its ability to retain moisture. This in turn is dictated by its total effective pore volume and its pore size distribution. The former controls the amount of water that may be held within the material while the latter controls the water and the space inside the material that is available for colonisation.

Under tropical climatic conditions, if the surfaces are warm with sufficient light, algae will develop in the water film on the surface, typically producing a bright green colouration, although sometimes dark green, brown and pink colourations occur. Algae often colonise a surface within one or two hours of rainfall, but the algal colouration disappears just as rapidly as the surface dries. Most of the algae spores are killed upon drying out but the minute quantities that remain are enough to redevelop and multiply when dampness returns. Organic deposits on the surface also encourage biological growth.

They derive their energy largely from sunlight but need liquid water for survival. In some materials, salts that are present inherently in the material serve as a breeding ground for these growths.

This chapter discusses seven common types of facade material with respect to their susceptibility to staining. The materials are:

- *Exposed brickwork.*
- *Concrete.*
- *Natural Stones.*
- *Tiles.*
- *Metal.*
- *Glazing.*
- *Plaster and paint.*

3.2 Exposed Brick

One of the oldest building material, brick, continues to be a most popular and leading construction material because it is cheap, durable and easy to handle and work with. Bricks are commonly manufactured in three forms; common bricks, facing bricks and engineering bricks. Common bricks are general-purpose bricks used for filling, backing and in walls where appearance is of no consequence. Facing bricks however are used on walls to create a pleasing appearance [5]. They are more durable under severe exposure and have a better appearance in terms of texture and colour than common bricks. Engineering bricks are stronger, harder, impermeable and have a smoother surface texture than common and facing bricks. In Singapore, the use of facing bricks as a facade material is most frequently seen in low-rise private residential houses as well as high-rise public flats (Fig. 3.2). The material characteristics of bricks in relation to staining problems are shown in Table 3.1.

Figures 3.5 and 3.6 show efflorescence on brick walls resulting from salts dissolution. Efflorescence commonly occurs on porous wall materials such as bricks, in particular clay bricks. Clay bricks contain sulphates, which dissolve readily in the water when the bricks become wet after a rainfall. The soluble salts are transported to the surface, where upon evaporation they crystallise to become white deposits. They

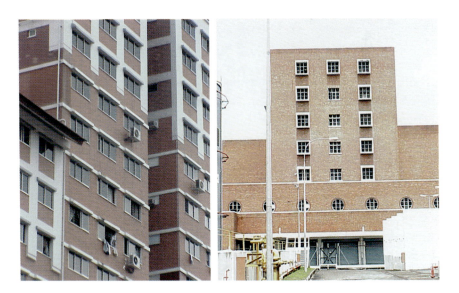

Figure 3.2. Use of facing bricks in high-rise buildings.

Table 3.1. Characteristics relating to the staining problems of brick facades.

Bricks Properties	Lightweight	Common and Fair Face	Engineering	Sulphate Resistant
Porosity (%) (volume of pores/total vol. Of whole material)	Porosity of brick is attributed to its fine capillaries. By virtue of capillary effect, the rate of moisture transport in the brick is ten times faster than in other building materials.			
Water Absorption [(saturation wt.-dry wt.) /dry wt. Of material]	12–14%	8–14%	6–7%	7–8%
Surface Texture	Mortar joints and coarse surfaces of brick create depressions and channels affect runoff flow (Fig. 3.3).			
Colour	Uniform deep red or cherry colour (due to ferric oxide). Impurities in clay can produce greenish-yellow/ buff coloured bricks (Fig. 3.4) [6]. Brick colour is permanent and will not fade because of weathering.			
Resistance to Chemical Attacks	Clay in bricks makes it susceptible to chemical attacks due to the presence of sulphates of calcium, magnesium, sodium and potassium. Rainwater dissolves calcium salts in bricks to produce efflorescence.			
Resistance to Biological Growth	Inherent salts, water absorption and retention in bricks support biological growth.			
Relevant Standards	SS CP67 Part 1: 1997, BS 3921: 1985, BS 5628:3: 2001, BS 6270-3:1991, ASTM C216-02, ASTM C1400-01, ASTM C1403-00			

usually disfigure the walls temporarily and will disappear after prolonged weathering. However, continuing efflorescence may indicate that water penetration into the wall is occurring.

Brick walls may also be susceptible to lime staining. Although lime stains have similar appearances as efflorescence, they are harder to remove. Lime stains form when free lime from the mortar joints or from concrete components in close proximity is leached out by water to the surface. Upon exposure to carbon dioxide, it is converted to calcium carbonate, which is almost insoluble in water. Preventing lime staining requires the exclusion of water from wall surfaces by having good water shedding details. Figures 3.7 and 3.8 show lime staining on brick walls.

Figure 3.3. Rough surface texture of brick walls.

Figure 3.4. Impurities in clay can produce greenish-yellow/buff coloured bricks.

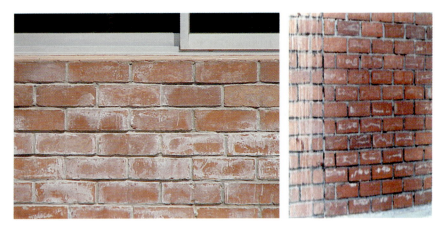

Figure 3.5. Staining due to efflorescence.

Figure 3.6. Staining due to efflorescence.

Figure 3.7. Staining due to leaching of salts from mortar joints.

Figure 3.8. Localised occurrence of salts leaching from mortar joints.

Figure 3.9. The side of the facade that is affected by biological staining.

Staining on brick walls may also be in the form of biological growth. Biological growth such as algae growth on brickwalls is common due to its ability to retain moisture to remain damp for a long period of time after rainfalls (Fig. 3.9). The moisture film allows dirt to be adhered easily, further promoting growth.

The rough surface texture of brickwalls also retains and traps dirt particles more easily. Figure 3.10 shows dirt streaking on a painted brick wall. The surface irregularities on the wall, as well as the window ledge concentrate runoff in discrete streams. This causes stain streaks to build up at areas where runoff flow is usually directed to run over.

3.3 Concrete

Concrete facades can either be cast *in situ* or prefabricated in the yard before transported to site to be installed as precast concrete components.

Figure 3.10. Staining due to dirt.

Precast concrete components came into use as cladding material during the development of mass high-rise housing. Advantages from using prefabricated components include savings from economies of scale, repetitive production leading to familiarity with processes and reduction in mistakes and savings in construction time. In Singapore, they are used as cladding materials either with finishes cast together in the mould or finished with tiles or stone facings. With the strong emphasis placed on buildability and maintainability of buildings, the use of precast concrete components is expected to increase tremendously. This has been evident with the widespread use of prefabricated concrete facade panels on public residential housing in Singapore (Fig. 3.11). The material characteristics of concrete in relation to staining problems are shown in Table 3.2.

Figure 3.14 shows large patches of stains on a concrete facade. This staining is due to the uneven washing off of dirt during rain. Dust accumulated on the facade is washed down during rainfall. If rainfall is

Figure 3.11. Precast concrete facades.

Table 3.2. Characteristics relating to the staining problems of concrete facades.

Concrete Properties	Concrete
Porosity (%) (volume of pores/total vol. of whole material)	Water/cement ratio determines the porosity of hardened cement paste at a given stage of hydration (Fig. 3.12) [7]. It generally exceeds 15%.
Water Absorption [(saturation wt.-dry wt.) /dry wt. of material]	Dependent on air voids (1–3% of volume). Moisture causes blotches/streaking after rainfall [8,9]. Moisture penetration may corrode steel reinforcement in concrete to form rust stains.
Surface Texture	Smooth to rough depending on aggregates and face moulds used for casting. Textured and profile surfaces mask stains better (Fig. 3.13) [7].
Colour	Wide-ranging colours depending on colour added to matrix. Coats of paint can be given to achieve the desired colour [7– 8].
Resistance to Chemical Attacks	Dependent on quality of concrete. Rainwater dissolves calcium salts in concrete to produce efflorescence and lime staining.
Resistance to Biological Growth	Damp concrete surface traps dirt and promotes biological growth.
Relevant Standards	BS 6270-3:1991, BS 5628:3: 2001, AS HB 84:1996

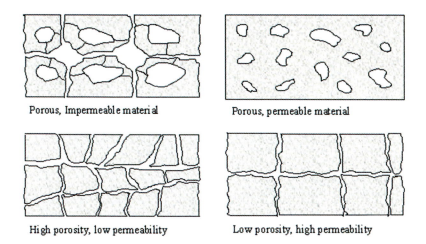

Porous, Impermeable material   Porous, permeable material

High porosity, low permeability   Low porosity, high permeability

Figure 3.12. Concrete structures with different porosity and permeability (Source: Concrete Society Working Party, Permeability testing of site concrete – a review of methods and experience. Final draft, Nov. 1985, Permeability of concrete and its control, Papers, London, 12 December 1985, pp. 1–68).

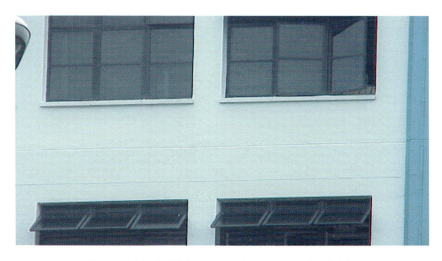

Figure 3.13. Prefabrication results in a smoother finish.

Figure 3.14. Staining on concrete facade.

weak, the washing effect is not thorough. Concrete, being porous, readily absorbs moisture during rain through capillary sorption. Upon saturation of the concrete wall, a film of water laden with dirt will form on the wall surface. As water dries up, dirt stays on the facade, forming dark stains. Staining on concrete facade can also be due to the growth of algae (Fig. 3.15).

3.4 Natural Stones

Natural stones are formed naturally under different conditions and hence no two pieces bear the same composition and appearance. A good quality natural stone can be aesthetically pleasing and durable when used as a facade material. Table 3.3 shows a checklist for assessing the quality of natural stones. Some examples of commonly used natural stones in Singapore are limestone, sandstone, marble, slate, quartzite and granite, with granite being the most commonly used stone facade material.

Granite is a crystalline igneous rock containing mainly quartz and feldspar. Granite is very hard and heavy and it is one of the most durable

Figure 3.15. Algae growth on concrete facade.

materials used for external cladding (Fig. 3.16). It can be expressed in many finishes ranging from polished to rough. Granite tiles are usually installed on the tile bed in large pieces using mechanical means to reduce the incidence of debonding [10, 11]. The cost of granite may be initially high but the product is long-lasting and requires little cleaning.

Granite is suitable for use on buildings facing the sea. They are not affected by the weathering action of sand particles blown by wind. Granite is also acid fumed and smoke-proof and is also suitable for use on buildings in areas where pollution levels are high.

Figure 3.19 illustrates examples of staining problems on natural stone surfaces. The types of staining as observed in both figures can be defined as residue rundown. Due to the porous nature of natural stone, dirt residue that is washed down might get collected in the pores of the stone, and upon drying, dark stains are formed. Table 3.4 explains the characteristics of stones in relation to staining problems.

All stones absorb water. However, its water absorptivity may reduce upon sealing and polishing. The absorption of runoff may occur directly on the surface of the facade or through expansion joints and then

Table 3.3. Checklist for a good quality facade stone.

Quality	Performance
Appearance	Fine and compact texture. Veining and colour variation should not be excessive as they indicate softness.
Structure	Structure of stone should be homogenous without presence of voids, cracks or patches of loose or soft aggregates. The cross section of the stone should not be dull in appearance.
Weight	Indication of porosity and density. The heavier the stone, the lower is its porosity, the denser is the stone structure and hence the higher its stability.
Toughness	Toughness is the measure of impact that the stone can take. The stone used should be relatively tough to withstand impacts and high wind loads.
Hardness	More important for floor tiles than facade tiles.
Permeability	Permeability should be low so that the degree and depth of moisture penetration into the stone is low. Permissible water absorption is given in Table 3.4.
Absorptivity	Honed and textured stones have more open pores on its surface than polished stones and hence absortivity is higher and staining is more likely.
Weatherability	Stone should resist against wear and tear due to weather elements. When exposed to sunlight, it should not fade. Dark coloured stones tend to fade more easily.
	It should not disintegrate when subjected to thermal shocks and chemical attacks. Marble slabs show a distinct distortion when subjected to heat. When the stone is exposed to uneven heating, the side that is heated may expand and warp. On cooling, the slab may not return to its original shape.
Workability	Fabricating qualities should be consistently good so that cutting, and dressing to the required shape and size is economical.

penetrate into the granite tile. It is hence important to seal all six sides of the stone panel, but at the same time ensuring breatheability. The large patches of staining shown in Figs. 3.20 and 3.21 are due to the effect of dampness on the minerals of porous granite.

In Fig. 3.22, staining is shown to have arisen from lime leaching. Being a permeable material, natural stones allow water to seep into the backing wall. Water may also reach the substrate by penetrating through

Table 3.4. Characteristics relating to the staining problems of natural stone facades.

Natural Stone / Properties	Granite	Marble	Sandstone	Limestone
Porosity (%) (volume of pores/total vol. of whole material)	0.4–1.5%	0.6–2.3%	0.5–35.0%	0.6–31.0%
Water Absorption [(saturation wt.-dry wt.) /dry wt. of material]	After sealing stones with impregnator, water adsorption values may be reduced. In its sealed and polished form, runoff generates quickly to flow on the stone facade in discrete streams. However, the stone may still be absorbent even after polishing and sealing.			
	0.2–0.5%	0.2–0.6%	0.2–9%	0.2–12%
Surface Texture	Polish to give fine and evenly granular texture. Flame to give rough texture (Fig. 3.17).	Takes fine polish. Requires regular re-polishing.	Weathers well when free from lime and iron. Due to high porosity, requires sealing to prevent dirt penetration into pores.	Due to high porosity, requires sealing to prevent dirt penetration into pores. Surface will weather irregularly due to mineral composition, resulting in roughness.
Colour	Stone colour may change by applying chemical dyes.			
	Wide range (depends upon colour of feldspar) (Fig. 3.18) [12]. Usually dark pink or deep red. Granites with red and black minerals may fade.	Wide range due to the variability of accessory minerals.	White to buff. Full range of earth colours. Iron oxide present in sandstone may develop rust spots upon oxidation.	Buff, gray. Limestone remains clean in portions facing rain but retains a film of soot in sheltered areas. This results in a strong colour contrast. Yellows after long exposure to atmosphere [12].
Resistance to Chemical Attacks	Good resistance against acid. May produce rust spots if it contains ferrous mineral compounds.	Poor resistance against acid.	Good resistance against acid.	Poor resistance against acid.
Resistance to Biological Growth	Does not promote biological growth. Stones with large pore size may be attacked more easily.			
Relevant Standards	BS 8221-1:2000, BS 8221-2:2000, BS 8298:1994, ASTM C1248-93 (1998), ASTM C1496-01, SS CP 67, BS ISO 11600: 2002, DIN 18540			

Figure 3.16. Granite facade.

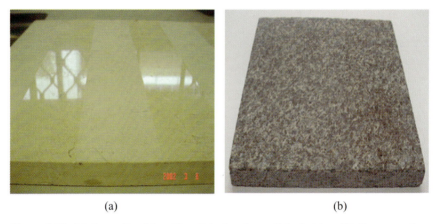

(a)   (b)

Figure 3.17. (a) A granite slab with polished and unpolished strips. (b) Coarse surface texture of flamed granite.

cracks and mortar joints. Water dissolves lime and other soluble salts from the concrete backing and leaches out the lime and other soluble salts from the concrete backing. The leaching of these compounds to the facade surface would form white stains on the facade upon drying. Lime

Figure 3.18. Different colours of granite.

Figure 3.19. Staining arising from residue rundown.

Figure 3.20. Wetting pattern after a rainfall.

Figure 3.21. Rising dampness causes staining at ground level.

Figure 3.22. Lime leaching from mortar joints or from concrete backing wall.

leaching is more serious along the horizontal joints since water enters and exits more easily than at vertical joints.

3.5 Tiles

Although the use of tiles as an external wall finish has reduced in recent years, there is still a large number of buildings that had been clad with tiles [13–15]. Common tiles that were used for cladding facades include mosaic and ceramic tiles (Figs. 3.23 and 3.24). Their material characteristics with regard to staining problems are summarised in Table 3.5 [16].

Figures 3.26 and 3.27 illustrate examples of staining arising from localised dirt deposition by rainwater on tiles. The delamination of small dimension tiles in Fig. 3.26 has resulted in voids or crevices, which trap dirt and dust to cause staining during rainfalls [17, 18]. Mortar infill at joints has a higher water absorption value than tiles. This allows dirt to adhere more easily to the damp surfaces. Where tile dimension is small, the wall area that is made up by mortar joints will be large and staining from the joints will be more visible.

Figure 3.23. Variety of colours for ceramic tile.

Figure 3.24. Mosaic tiles as an external wall finish.

Figure 3.28 shows cases where severe lime leaching has taken place on tiled facades. In these cases, the mechanism of lime leaching is almost similar. Rainwater that seeps through cracks, chip-offs or mortar joints of the tiled wall into the cementitious backing wall or substrate will leach out the soluble salts. The penetrated water will emerge again at joints, crack lines or chip-offs and upon evaporation, crystallises to form white deposits. The material properties of tiles do not contribute to the staining in such cases. As compared to efflorescence, lime leaching takes a longer time to form but once formed, it is harder to remove. Rectification should involve the prevention of water flow over the facade.

Table 3.5. Characteristics relating to the staining problems of tiled facades.

Tile / Properties	Ceramic Tile	Homogenous	Mosaic tile		
			Non-Vitreous	Vitreous	Impervious
Porosity (%) (volume of pores/total vol. of whole material)	0–70% depending on glazed/ unglazed.	Almost zero	Up to 30%	> 3%	Almost zero
Water Absorption [(saturation wt.- dry wt.) /dry wt. of material]	3–6%	0.04–0.5%	> 7%	0.5–3%	< 0.5%
Surface Texture	Glazed tiles are smooth surfaced unlike unglazed tiles. The smaller the dimension of the tiles, the more mortar joints there are and the overall surface will be rough and irregular.				
Colour	Glazed layer gives tile its colour.	Colour of mosaic tiles goes all the way to tile. Mixing of tile colours offer visual aesthetics.			
Resistance to Chemical Attacks	Inert to chemical attacks. Often stained by lime leaching out from mortar joints or backing wall.				
Resistance to Biological Growth	More likely to occur on joints between tiles (Fig. 3.25).				
Relevant Standards	BS 5385-2: 1991, ASTM D5343-97, ASTM C482-02 (1996), ASTM C650-97, ASTM C1378-97, SS CP68: 1997, AS 4459.2-99, AS 4459.3-99, ISO 10545-3: 1995, ISO 10545-14: 1995				

Figure 3.25. Algae growth at mortar joints of ceramic tiles.

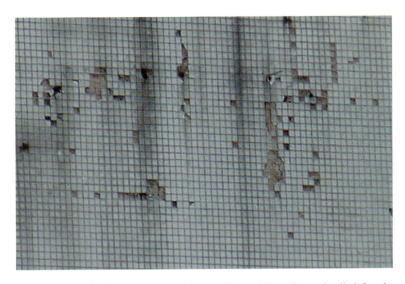

Figure 3.26. Debonding accompanied with dirt staining of mosaic tiled facade.

Figure 3.27. Staining of tiled facade.

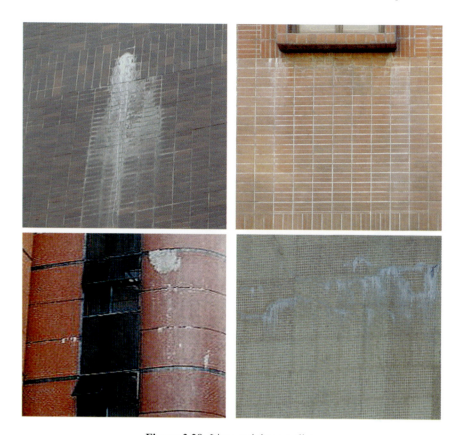

Figure 3.28. Lime staining on tiles.

3.6 Metal

Metal cladding has gained huge popularity in Singpore over the last two decades. Besides new constructions, numerous old building facades have also been over-clad or re-clad with metal cladding systems involving aluminum and glass. This has been seen to be a good alternative in building refurbishment (Fig. 3.29). The most common type of metal used in cladding systems of commercial and industrial buildings in Singapore is aluminum and steel (Fig. 3.30). The characteristics of aluminum and steel are compiled in Table 3.6.

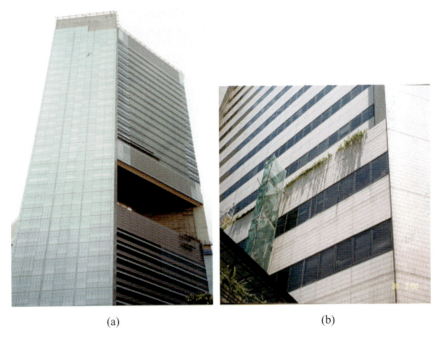

(a)   (b)

Figure 3.29. Aluminum and glass over-cladding (a) over the old tiled facade (b).

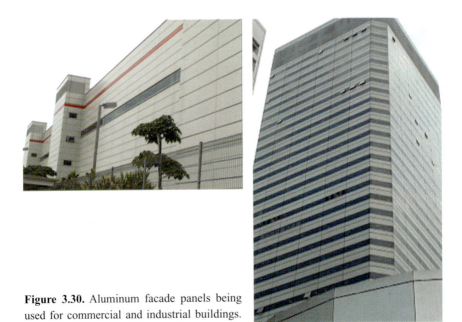

Figure 3.30. Aluminum facade panels being used for commercial and industrial buildings.

Table 3.6. Characteristics relating to the staining problems of metal facades.

Metal Properties	Steel	Aluminium	
		Fluorocarbon (Polyvinyl Fluoride)	Anodised Aluminum (Electro Deposited Oxide)
Porosity (%) (volume of pores/total vol. of whole material)	Low porosity due to the close and regular arrangement of atoms in structure. Due to its hydrophobic nature, rainwater forms discrete streams on its surface (Fig 3.31).		
Water Absorption [(saturation wt.-dry wt.) /dry wt. of material]	~ 0. Runoff is generated almost immediately during a rainfall. Heavy rainfalls can provide a strong washing effect to cleanse the facade of general dirt.		
Surface Texture	Smooth. Hydrophilic nature causes a thin layer of moisture to form on metal. This moisture layer attracts and retains atmospheric dirt.		
	Grained or smooth.	Stucco, grained or smooth.	
Colour	Generally large colour range (Fig. 3.32). Light colour shows off stains. Colour loss more likely for bright colours. Dark coloured panels absorb heat better and experience greater expansion.		
	Large range of colours including exotic colours.	Anodised or with paint applied coat (PVDF).	
		Unlimited range. Excellent panel-to-panel match.	Limited. Noticeable colour contrast between panel-to-panel.
Resistance to Chemical Attacks	Can tarnish or corrode under ultraviolet radiation. Requires frequent maintenance to retain shine.	Coating usually applied to protect base metal and increase durability. Scratches/abrasions may be inflicted during maintenance works, which may damage coating and expose the base metal.	
		Excellent resistance to pollution and salt sprays. Will not be stained by mortar. Durability of 20 yrs.	Stains more easily. Mortar stains surface. Whitening on salt sprays. Durability of 15 yrs.
Resistance to Biological Growth	Thin film of water traps dirt and organic particles. Organic particles may also be present in coating to support biological growth. Less prone to biological attacks than masonry walls.		
Relevant Standards	ASTM D523-99, ASTM D2244-02, AS 1562.1 1992, BS 6496:1984, BS 6213:2000		

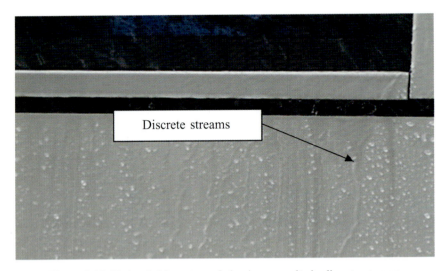

Figure 3.31. Hydrophobic nature of aluminum results in discrete streams.

Figure 3.32. Aluminum panels can be fabricated in many colours.

The main characteristics of metal cladding are:
- High strength but light-weight.
- Non-combustible.
- Weather resistant.
- Ductile.
- Offers a wide variety of trapezoidal profiles and coloured finishes.

Stainless steel, bronze and copper have also been widely used as cladding materials in other countries [19]. Aluminum cladding has overtaken steel in terms of popularity due to its strong, light, stable and adaptive qualities. With the right finish, aluminum will retain its appearance for many years even with little maintenance. Aluminum can be finished by a number of methods. Due to the Singapore's tropical environment, aluminum is usually finished with hardwearing fluorocarbon coat (PVDF). Anodising — a process whereby a hard patina that prevents further deterioration of the surface is produced — can also be used.

During heavy rainfalls, rainwater that impacts on the facade will generate runoff to bring about an efficient washing effect. However, the washing effect during a light rainfall is weak and may result in localised staining arising from dirt absorption, erosion and deposition. The degree of staining depends on the facade design and wetability of the facade material. Aluminum has an absorption coefficient of almost zero.

Figure 3.33 shows the vertical streaks of stains below the whole length of a ledge. This is due to airborne dirt accumulating on the

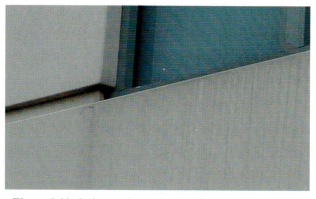

Figure 3.33. Stain streaks originating from a window ledge.

horizontal window ledge and later washed down by the runoff. These streaks are more visible against the light coloured panels. Figure 3.34 shows more severe dirt staining problems on fluorocarbon coated aluminum cladding. Long-term exposure to polluted air and chemical rain coupled with little or no maintenance has allowed the stains to build up. Chemical attack from cleaning solutions or abrasions from the use of wrong cleaning agents and equipment may also cause deterioration of the surface.

Figures 3.35 and 3.36 show staining associated with recessed or protruding joints, detailing faults and other factors. Chemical rain brought about by highly polluted air in urban and industralised areas is one of the causes of staining. Erosion of the facade over time due to chemical rain has degraded the facade and weakened the coating's resistance to chemical attacks, thus making it easier to attract dirt. Deterioration of sealant between metal panels can also cause staining (Fig. 3.37).

Biological growth such as algae and moss may also stain metallic facades. Algaes thrive on the thin film of moisture on the panels. Their growth may be supported by the presence of organic particles in the

Figure 3.34. Dirt stains on light coloured aluminum panels show off even more.

Figure 3.35. Staining on enamel steel

Figure 3.36. Staining on enamel steel resulting in corrosion of panels.

Figure 3.37. Sealant failure resulting in black stains.

Figure 3.38. Biological growth on facade at ground level and in proximity to vegetation.

atmosphere or by the organic substances present in the fluorocarbon coated metal panels (Figs. 3.38 and 3.39).

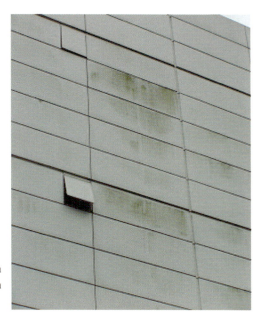

Figure 3.39. Biological growth on facade panels at 4th storey and in proximity to trees.

3.7 Glazing

The use of large glazing panels on facades has become widespread due to improvements in technologies which enable glass to span many floors but yet withstand high wind loadings, and also due to building owners becoming more willing to spend lavishly for higher aesthetical qualities for their buildings (Fig. 3.40). Glass is usually used at vision areas as infill materials for curtain wall systems.

The material properties of glass with respect to staining problems are summarised in Table 3.7.

Figure 3.43 shows dark streaks below the joints. The open joints coupled with glass's hydrophilic nature trap dirt. When rainwater runoff flows over the joints, the dirt is washed out to stain the glass. Like metal, glass has an absorption coefficient of zero. Hence, during heavy rainfall, impacting rainwater and its runoff is able to effectively wash off general dirt stains from glazed surfaces. This allows glass to have a "self-cleansing effect". However, when the rainfall is light and intermittent, the runoff generated flows in discrete streams and does not provide an efficient washing effect. This results in dirt streaks forming almost immediately. Over time, the aesthetic quality of the glazed facade is affected. Intermediate roofs or projections without adequate drainage system or features to "throw out" the rain water, will result in accumulated dirt on the roof surface, and this will be washed down onto the facade surface, resulting in obvious streaking (Fig. 3.44). The pattern of the streaking is determined by the rain flow pattern, which in turn, is determined by the design profile, material, wind, etc. (see Chapter 4).

In cases where sealant is used between glass panels, its chemical make-up and electrostatic nature may attract and retain dirt. When rain water runs over the sealants, loosely held dirt particles will be dislodged and carried downwards to form stain streaks that originates from the

Figure 3.40. Buildings clad entirely in glass exude modernity.

Table 3.7. Characteristics relating to the staining problems of glazed facades.

Glass Properties	Glass
Porosity (%) (volume of pores/total vol. of whole material)	Not applicable.
Water Absorption [(saturation wt.- dry wt.) /dry wt. of material]	~ 0. All of the rainwater forms runoff. Hydrophobic property results in discrete streams and prevents even washing. May result in non-uniformly stained walls on wall areas directly below glazed areas (Fig 3.41). Walls below glazing should be materials with colour or texture which can mask the stain streaks [1].
Surface Texture	Very smooth. Microscopic cracks exist on surface. Hydophilic nature causes a thin moisture to form on glass.
Colour	Can be coated. Usually gray, bronze or green. Coating should be balanced with daylight transmission and views to outside. Dark coats can mask stains (Fig. 3.42).
Resistance to Chemical Attacks	Affected by lime, caustic soda, ammonia and strong acids usually from cleaning solutions. Airborne acids, alkali and solvents from industrial sources may damage the coating of glass [20].
Resistance to Biological Growth	Inert to attacks. But film of moisture may promote growth.
Relevant Standards	BS 952 –1:1995, BS 4254:1983, BS 5516:1991, BS 6262:1982, BS 5889:1989, ASTM C719-98, ASTMC603-97, ASTM C510-90, ASTM C792-98, ASTM C793-02, ASTM C1087-00, ASTM C1193-00.

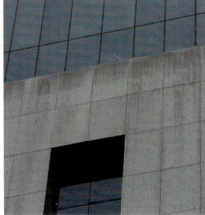

Figure 3.41. Staining of stone walls due to water running off from glass.

Figure 3.42. Glazed facade with different tints and reflectivity.

Figure 3.43. Staining on glass originating from open joints.

sealant joint. Disintegration of sealants worsens the staining problem (Fig. 3.45). Exposure to heat and moist conditions in the form of rain, dew and high humidity has caused many sealants to fail prematurely. The hydrolysis of the polymer chemically produces hydroxyl radicals in the presence of sunlight. This product is detrimental to the structure of sealants and may result in leaching of plasticisers and stabilisers [14]. This may result in the upward migration of plasticisers and stabilisers in the sealant as shown in Fig. 3.46. Figure 3.47 shows staining on glazed surfaces due to the deterioration of sealant.

Figure 3.44. Accumulation and washing off of dirt from the sloping roof causes staining on the glass facade.

Figure 3.45. Sealant between the glass panels retains dirt and causes staining to be more severe.

Figure 3.46. Upward migration of plasticisers and stabilisers in the sealant causes the stains shown.

Figure 3.47. Staining of glass due to corrosion of sealant.

Figure 3.48. Staining of glass due to corrosion of transoms and mullions

In some cases, corrosion of other components that make up the facade may cause staining to glass (Figs. 3.48 and 3.49). When cementitious adhesives are used in a glass facade, rainwater hitting the facade may carry the soluble salts in the adhesive to the surface, which will form efflorescence on drying.

Facades that are exposed to direct sunlight, prolonged rainfalls and in close proximity to vegetation may be prone to algae attack (Fig. 3.50).

3.8 Plaster and Paint Surface Coatings

Two common surface coatings used in the region are plaster and paint. The characteristics of plaster and paint are summarised in Tables 3.8 and 3.9 respectively. Plaster and paint coatings serve as decorative, functional and protective coverings which upon spreading over a surface, will dry into a continuous solid film. Plaster and paint can come in different colours and textures. Table 3.10 shows some examples of various types of finishes, smooth or textured, that is available with plaster.

Figure 3.49. Staining from weepholes results in a regular staining pattern throughout the facade.

Figure 3.50. Algae growth on glass.

Table 3.8. Characteristics relating to the staining problems of exposed plaster facades.

Exposed Plaster Properties	Textured Finish	Plain Finish
Porosity (%) (volume of pores/total vol. of whole material)	Depends on size and proportion of aggregates.	
Water Absorption [(saturation wt.-dry wt.) /dry wt. of material]	Determined by the water/plaster ratio of the mixes [21]. If plaster is made denser, the water absorption level will reduce.	
	Flow of rainwater over the surface is distributed. Staining is thus more distributed.	Flow of rainwater over surface is concentrated. Staining is thus more concentrated.
Surface Texture	Traps and retains dirt easily. Textured and rough surface may mask stains.	Stain streaks may show easily.
Colour	Different colours can be obtained by adding colour pigment to mix. Light colour shows off stains.	
	Uniform appearance for paints can be easily achieved.	Uniform appearance for paint may be difficult to achieve.
Resistance to Chemical Attacks	Prone to chemical attacks due to aggregates used. Gypsum plaster may suffer attacks from Portland cement due to formation of calcium salts [22].	
Resistance to Biological Growth	Water may get trapped in the plaster layer and accelerate biological growth. Organic material present in the plaster mix may encourage biological growth.	Prone to biological attacks, but less than textured surfaces. Constituents in plaster mix may encourage biological growth.
Relevant Standards	BS 8000-10:1995, BS 5262:1991, BS 1191-2:1973, BS EN1015-19:1999, BS 00/106850 DC prEN 13914-1, BS 00/103730 DC prEN 998-1, ASTM C897-00, ASTM C932-98a	

Although many coatings are able to "breathe" out any moisture trapped beneath the coating, the rate of evaporation may not be fast enough and this provides a breeding ground for algae and fungi. When the colour of plaster or paint is light, staining would be readily visible.

Figures 3.51 and 3.52 show dark stains on plaster-and-paint surfaces arising from dirt deposition. The textured surfaces of plastered walls are capable of trapping and retaining dirt particles. When runoff flows over

Table 3.9. Characteristics relating to the staining problems of painted facades.

Paint Systems on Plastered, Concrete or Brick Walls / Properties	Acrylic Co-Polymer Emulsion	Veova (Modified Acrylic) Emulsion	Epoxy	Acrylic Emulsion	Solvent-Based Acrylic	High Performance Acrylic Emulsion	2-Pack Poly-Urethane	Texture Coating Water-Based Acrylic Topcoat	Texture Coating Poly-Urethane Topcoat
Porosity (%) (volume of pores/total vol. of whole material)	Depends on overall system.								
Water absorption [(saturation wt.-dry wt.)/dry wt. of material]	Depends on overall system.								
Surface Texture	Textured.	Luxurious sheen emulsion that gives a smooth and silky appearance.	Textured.	Decorative, high build, textured, non-gloss.	Satin finish.	Smooth.	Various textures may be created with different types of brushes.	Textured.	Textured.
Colour	Matt.	Sheen.	Low gloss/gloss.	Matt.	Semi-gloss.	Low sheen.	Good gloss retention.	Semi-gloss.	Gloss.
Resistance to Chemical Attacks	Good.	Very good.	Excellent.	Good.	Very good. Good durability to weather.	Very good.	Excellent. Good weathering resistance.	Very good.	Excellent.
Resistance to Biological Growth	Good.	Good.	Excellent.	Good.	Very good.	Excellent.	Excellent.	Very good.	Excellent.
Relevant Standards	SS CP 22:1981, SS 5, SS 7:1998, SS 34:1998, SS 150:1998, SS 345:1990, Good Industry Practices- Painting (BCA)								

Figure 3.51. Dirt staining of Shanghai plaster.

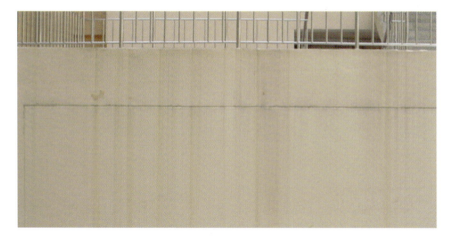

Figure 3.52. Dirt staining on a smooth plaster-and-paint wall.

the surface, it flows along depressions and crevices of the rough surface. Any dirt that is picked up along its path is redeposited. The flow pattern causes the resultant dirt stains to be distributed. On smooth plaster surfaces, staining may also occur but the staining pattern is likely to be less concentrated as dirt particles are washed along in more even and

Table 3.10. Types of plaster finish.

1) Smooth

The final coat is finished with wooden trowel or steel trowel and is made rather flat and smooth. In Singapore, most of plain finishes are painted.

2) Textured

Apply scraping or rubbing on 2nd coat when it is still not completely hardened by using hand tools or trowels.

3) Rough cast/pebble dash

Consists of 2 coats. The final coat is applied by throwing small pebbles or other coarse aggregates onto the 1st coat and left untrowelled. The coarseness of the texture depends upon the size and shape of the coarse aggregate used.

4) Shanghai plaster

Coarse aggregates of small pebbles are added to the mix of the final coat to achieve a rough textured finish. Before the surface hardens, it is scrubbed with fibre brushes to expose the aggregates and then rinsed off with water.

Table 3.10. (*Continued*)

5) Scraped finish

Scraped finish is a textured finish. The final coat of mortar is allowed to harden for several hours and the surface is then scraped with a suitable tool to remove the outer surface and some of the coarser particles.

6) Spattered finish

The spattered finish is produced by a hand-operated machine which flicks droplets of a workable mortar onto the wall where they set and harden.

7) Machine sprayed finish

Power-operated machine can be used to apply the plaster.

8) Rubbed machine sprayed finish

Some days after a plaster has been sprayed, it may be rubbed with a tool to produce a smooth flat outer surface with remaining texture.

concentrated flows. The washing effect on a smooth wall is greater and more effective in times of heavier rainfall.

Certain organic materials that support biological growth may be present in the aggregates that make up plaster. The use of such composition may result in biological staining (Figs. 3.53 and 3.54).

Painting is commonly done on plaster walls to enhance its aesthetic qualities. Paints are relatively permeable to liquids and gases. Depending on the type of paint system used, the water absorptivity and chemical resistance of the overall wall system can be improved. Table 3.9 shows the various characteristics of different paint systems with respect to staining [23, 24].

Figure 3.53. Biological staining on plaster walls.

Figure 3.54. Staining due to algae infestation.

3.9 Material Interfaces

Facades are usually made up of more than one material either because different materials are required for their characteristic functions or a combination is neccesary to increase the aesthetic qualities of the facade. The use of different materials on a facade, whether necessary or not can lead to visual changes in the form of stains, if the materials are not chemically compatible. Therefore, besides understanding the characteristics properties of individual materials, it may also be important for the designer to know the effects due to the interaction of different materials when put together. Table 3.11 summarises the interactions between different materials when they are used in combination [6].

3.9.1. Bricks — Concrete/Plaster

Both concrete/plaster and brick mortar contains calcium hydroxide and alkaline salts. Calcium hydroxide efflorescence on the surface reacts with carbon dioxide to give insoluble calcium carbonate, which causes white encrustations on the adjacent material as well as on the material itself. Alkali salts found in cement also causes efflorescence. Bricks that exude sulphates should not be located against concrete or mortar which is susceptible to sulphate attack.

3.9.2. Bricks/Concrete/Plaster — Natural Stone

When large amounts of runoff flow over bricks, concrete or plastered surfaces, lime products from them or from their mortar joints may be deposited on the natural stone. Bricks contain certain salts, which may cause them to efflorescence but also discolour the natural stone.

3.9.3. Bricks/Concrete/Plaster — Glass

The surface of glass may be stained and damaged by the runoff carrying alkali silicates extracted when the runoff flows over brick mortar, fresh

Table 3.11. Material interfaces [6].

Facade material	Brick/Concrete/Plaster (above)	Natural Stone (above)	Glass (above)	Metals (above)	Paints (above)
Brick/Concrete /Plaster		No effect.	Runoff generated easily runs onto the bricks and causes bricks to remain damp for a long period of time. Implications for algae growth. Dirt streaks may form on bricks.	Runoff generated easily from metal surfaces runs over the bricks and results in clean strips on bricks below vertical member elements.	Chalk particles from paint chalking deposited on underlying materials are difficult to remove.
Natural Stone	Lime products from brick mortar can be deposited on natural stone. Brick contains salts which can cause efflorescence but also discolour the natural stone.		Runoff generated easily runs over natural stone and result in clean strips below vertical member elements.	Metals may produce fine oxidizing products which will be carried away by runoff and deposited on the natural stone.	Chalking of paint produces fine chalk particles which may stain natural stone.
Glass	Glass can be stained by alkali silicates extracted by runoff from brick mortar. A deposit of silica forms on the glass and is hard to remove. Calcium and alkali-containing materials are easier to remove.	Alkaline solutions of limestone may form a deposit on glass. Can be cleaned off easily.		Streaking due to deposition of dirt onto glass surfaces. Rust (iron oxide) stains will be difficult to remove. Metal oxides produced due to weathering and allowed to accumulate is hard to remove.	Chalking of paint produces fine chalk particles which may decrease visibility of vision glass.
Metals	Lime products from brick mortar can show white streaks on dark coloured metal surfaces.	Alkaline solutions of limestone may attack and cause deterioration to aluminium.	Dirt running off glass surfaces can soil light-coloured metal panels.		Chalking of paint produces fine chalk particles which may stain dark coloured metal surfaces.
Paints	Calcium hydrate products react with paint pigments causing it to discolourize. Alkali salts could cause peeling of paint coatings.	No effect.	Increased washing effect on painted surface. May cause discoloration.	Increased washing effect on painted surface. May cause discoloration.	

concrete or plaster. A deposit of silica may be formed on the glass surface and may be hard to remove. Besides silica, calcium and alkali-containing materials may also be deposited, though these are easier to be removed. Silicone sealants used on such walls may also discolour glass surfaces. On the other hand, introduction of a glazed opening often has the effect of concentrating surface rainfall runoff and this may affect the durability of the materials in the vicinity. Runoff generated even from light rainfalls may run onto the brick, concrete or plaster wall, causing the wall to remain in damp conditions for longer periods of time. This will lead to algae growth and water seepage related stains.

3.9.4. Bricks/Concrete/Plaster — Metals

Alkali solutions leached out from mortar may react and cause metal to deteriorate. Aluminum in particular may dissolve when attacked by alkali. Conversely, when metal cladding is located above brick, concrete or plaster wall, runoff that is generated easily from metal surfaces will run over the wall and result in clean strips on the wall below vertical member elements. Such areas receive heavier runoff flows and is washed over more frequently.

3.9.5. Bricks/Concrete/Plaster — Paint

Calcium hydrates from mortar may react with certain pigments in paints to cause discolouration. Alkali salts running constantly over painted surfaces could cause peeling of the paint coatings. Products from the deterioration of paint such as fine chalk products from paint chalking may be deposited on underlying materials. These particles are very difficult to remove.

3.9.6. Natural Stone — Glass

Various forms of silicates are found in sandstone. Under certain circumstances, they are soluble in water and segregate on the surface in the form of coatings on glass, which are insoluble in water and acids. Due to high absorption of some stones, they should be treated with an oil/water repellent to minimise staining especially when runoff from glazed surfaces can run over it and redeposit dirt onto the natural stones.

3.9.7. Natural Stone — Metals

Natural stone does not stain adjacent metallic materials. However, alkaline solutions of limestone may attack and cause deterioration to aluminum. On the other hand, metals may produce fine oxidising products which will be carried away by runoff and deposited on the natural stones.

3.9.8. Natural Stone — Paint

Natural stone does not affect paint. However, when stones are sealed, runoff generated may increase. This may accelerate paint deterioration. The chalking of paint to produce fine chalk products can stain natural stone located beneath it.

3.9.9. Glass — Metals

Some metals such as steels release oxides over time. These metal oxides should not be allowed to build up over glass as it results in a layer of deposit that is difficult to remove. Otherwise, both glass and metals generate runoff almost immediately during rainfalls and do not affect each other visibly.

3.9.10. Glass — Paint

These materials do not affect each other visibly, except where there are horizontal ledges. Rainfall may carry dirt that had accumulated on the ledges and wash it down the painted surfaces. This could result in visible stains.

References

[1] Concrete Society Working Party, "Permeability testing of site concrete — A review of methods and experience", Final draft, Nov. 1985, *Permeability of Concrete and its Control*, Papers, London, 12 December 1985, pp. 1–68, 1985.

[2] H. Parker, *Materials and Methods of Architectural Construction*, 3rd Edition, John Wiley & Sons, New York, 1958.

[3] P. Parnham, *Prevention of Premature Staining of New Buildings*, E. & F. N. Spon, London, 1997.

[4] O. Beijer, *Weathering on External Walls of Concrete,* Swedish Concrete Research Council, Swedish Cement and Concrete Research Institute, Stockholm. 1980.

[5] W. G. Foulks, *Historic Building Facades: The Manual for Maintenance and Rehabilitation,* John Wiley & Sons, New York, 1997.

[6] L. G. W. Verhoef, *Soiling and Cleaning of Building Facades*, Report of the Technical Committee 62 SCF, RILEM, Chapman and Hall, London, 1988.

[7] D. Campbell-Allen and H. Roper, *Concrete Structures: Materials, Maintenance and Repair*, John Wiley & Sons, New York, 1991.

[8] A. J. Brookes, *Cladding of Buildings*, 3rd Edition, E. & F.N. Spon, 1998.

[9] A. J. Brookes, *Building Envelope*, Butterworth Architecture, London, 1990.

[10] M. Y. L. Chew, "Study of adhesion failure of wall tiles", *Building & Environment*, UK, Vol. 27, No. 4, pp. 493–499, 1992.

[11] F. Nashed, *Time-Saver Details for Exterior Wall Design*, McGraw-Hill, New York, 1995.

[12] D. Richardson, "The staining of natural stone", *Construction Repair*, March/April 1993, pp. 15–17, 1993.

[13] T. K. Ong and P. J. Alum, "A study of performance of external wall tiling system in Singapore", in Proceedings [of the] *International Conference on Building Envelope Systems and Technology*, pp. 51–56, Singapore: Center for Continuing Education, 1994.

[14] M. Y. L. Chew, C. W. Wong and L. H. Kang, *Building Facades: A Guide to Common Defects in Tropical Climates*, World Scientific, 1998.

[15] M. Y. L Chew, "Efficient maintenance: Overcoming building defects and ensuring durability," *Conference on Building Safety*, The Asia Business Forum, Kuala Lumpur, 4 & 5 April 1994.

[16] C. Palmonari, *Ceramic Floor and Wall Tiles: Performance and Controversies*, Sassuolo (Modena): Edi. Cer. S.p.A - Sassuolo, 1989.

[17] M. Y. L. Chew, "Factors attecting tile adhesion for external cladding", *Construction and Building Metarical*, UK, Vol. 13, No. 5, pp. 293–296, 1999.

[18] M. Y. L. Chew, "Use of intra-red thesmography for assessing tile delamination on building facades", *Journal of Real Estate & Construction*, Vol. 8, pp. 64–72, 1998.

[19] H. Gage, *Guide to Exposed Concrete Finishes*, Architectural Press, London, 1974.

[20] J. S. Amstock, *Handbook of Glass in Construction*, McGraw- Hill, New York, 1997.

[21] J. B. Taylor, *Plastering*, 5th Edition, Longman Scientific & Technical, Harlow, Essex, 1990.

[22] G. D. Taylor, *Materials of Construction*, 2nd Edition, Construction Press, London, 1983.

[23] "Good practice guide for paint", Building Control Authority, Singapore, 2001.

[24] G. E. Weismantel, *Paint Handbook*, McGraw-Hill, New York, 1981.

CHAPTER 4

DESIGN ASPECTS

4.1. General

Staining can be described as a building defect that manifests itself as a change in colour that detracts from the desired appearance of the element [1, 2]. This problem arises from a lack of understanding of the interaction of materials and climate, the failure to ensure that the designer's intentions are achieved in the construction and lack of adequate, or any, maintenance [3].

Stains form from the workings of three mechanisms:

- Firstly, there must be surface flow of water (*runoff*).
- Secondly, the flow of runoff must bring along with it dirt particles that is being retained on the facade material.
- Lastly, after the runoff has reached its *limit of flow* and dried out, dirt particles that are not washed off will manifest as stains.

Each stream of runoff can be defined by its maximum length and width beyond which the runoff would be too slow in rate and too small in volume to continue its downward flow. This maximum length and width is known as the *limit of flow* for the runoff [4]. It is at this limit of flow that dirt particles carried along are re-deposited to form stains as shown in Fig. 4.1.

More importantly, facades stain because runoff is prevented from flowing controllably and evenly over the facade. Thus, dirt picked up in its path will be redeposited in concentration at some areas. Runoff is prevented from flowing evenly in a controlled manner and dirt is unable to build up evenly on a facade largely due to the design features found on

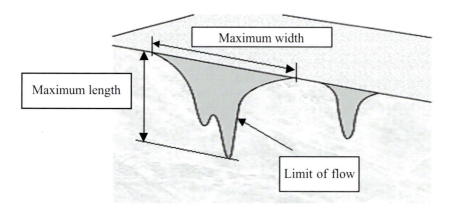

Figure 4.1. Dirt deposited at the runoff's limit of flow to form stains.

the facade [5–8]. On a planar and featureless facade such as one that is made up of similarly sized panels and does not have any recessions or protrusions, dirt build-up on the facade will be even and the facade will age evenly. It is therefore important to understand how design features affect the otherwise even runoff flow and dirt retention, and bring about stain marks.

With the development of new construction materials and technologies, bold and innovative design ideas for building facades have evolved. Although traditional facade materials such as bricks, stone, concrete and wood are still in use, many non-porous and less absorbent materials such as glass, plastic and metal are gaining popularity as they are able to offer greater flexibility in texture and colour. With new materials and more sophisticated facade designs, the provision of joints to effectively control rain runoff flow on the facade has to be re-looked. New facade designs that give more considerations to runoff flow needs to be advocated [9, 10].

The degree of staining is generally affected by the following factors:

• Atmospheric conditions, especially the pollution content of the air,
• The degree of exposure of the facade to atmospheric conditions,
• The flow pattern of rain water over the building facade,

- The type and property of the material used,
- The workmanship involved during the installation of the facade,
- The facade's design features.

Although atmospheric conditions are beyond our control, the designer nonetheless has to take the pollution level into consideration in the area where his building is sited when designing the facade and selecting facade materials [1, 4]. Heavy industrial areas experience higher pollution levels and as such, building components age and deteriorate faster [11, 12].

At a lower level, the different orientations of each building face means that the degree of exposure of each face to atmospheric agents differs. Facades that receive more rainfall and wind impact would be likely to experience greater dirt build-up and runoff flow and consequently, staining [13, 14].

Rain runoff flow pattern over a facade determines the degree of staining to a large extent. On many surfaces, this is a very thin layer, probably no more than a few tenths of a millimetre, and flows at low velocities of less than 1 m per minute. When rainwater flows over the building surface, dirt particles are picked up and redistributed to new positions, usually unevenly [5–8]. Furthermore, airborne dirt is also deposited on the building surface by rainwater. Thus, by designing the facade so that rain runoff can be controlled when it flows over the facade, dirt-streaking problems may be better addressed by designers.

Poor choice of facade materials and workmanship may lead to premature deterioration of building material [9, 10]. This is especially so if the facade is made up of many different types of materials with various properties. The deteriorating process or products from one material may stain the entire building facade, such as corrosion of metals and efflorescence of cementitious materials.

The rainwater flow pattern over a building facade is often disturbed by the facade's design features. Based on the case studies reported herein, it has been found that stains form most commonly around four main design features:

- Ledges.
- Joints misalignment.
- Protruding fixtures.
- Louver units.

The formation of stains is not irregular, but rather, is determinable by examining how runoff flow affects dirt build-up on the facade. Therefore, staining can be minimised or prevented by controlling the flow of runoff over design features.

4.2 Ledges

While a facade is generally flat and vertical, sometimes there are stretches of horizontal surfaces known as ledges that allow dirt particles of all sizes to be retained easily and runoff to flow over [15]. These horizontal surfaces can either protrude from or are recessed into the facade and they can range from less than a metre to several metres in length. Ledges are created on facades when window units are recessed or made to protrude from the facade; when access doors are incorporated in the facade; or simply when they are included as an architectural feature. Design features such as ledges are observed to cause disruption to the even flow of runoff. Should runoff be allowed to flow evenly over a featureless facade, the occurrence of dirt streaking may not be as intense since the even build-up of dirt would most likely be redistributed evenly over the facade [1, 4].

4.2.1. Factors Causing Staining around Ledges

Due to the downward flow of runoff, the area along and below the ledge is usually most prone to staining. Staining occurs at such areas due to:
- Presence of horizontal or inclined surfaces on the ledge where dirt particles of all sizes can settle and accumulate easily.
- The disruption of the otherwise even and regular flow of runoff as it flows over the ledge.

- Concentration of runoff at some areas on the ledge such that washing effect is irregular.

The top surface of a ledge, whether horizontal or inclined, allows dirt particles to settle and accumulate easily. This concentration of dirt on the ledge is washed off when runoff flows over it. The dirt particles will be carried along by the runoff as it flows over the ledge in a flow path determined by the profile of the ledge. The width of the ledge will affect the intensity of the stain streaks that originates from the ledge since a wider ledge will allow more dirt to settle on its surface and hence causes more pronounced stain streaks. However, a slight protrusion or recession from the general facade is adequate to retain dirt and cause staining (Figs. 4.2 and 4.3).

Runoff is diverted to concentrate in streams as it flows over the ledge. The horizontal plane of the ledge also causes the speed of the flow to be slowed down considerably, reducing the washing effect of the runoff.

For glazed curtain walls with transoms and mullions which are pronounced externally, staining patterns that follow that of ledges may also appear below the transom. This is because the actions of wind will drive flowing water down across the surface of the glass into concentrated streams that follow the line of these mullions. Stains may then be concentrated in the form of streaks on the walling below (Fig. 4.4).

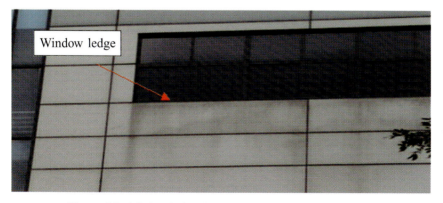

Figure 4.2. A ledge design feature that is experiencing staining.

Figure 4.3. Stain streaks due to ledges.

Figure 4.4. Diagrammatic sketch of water flowing over transom and mullions.

4.2.2. Staining Mechanisms around Ledges

Staining originates from the ledges as the runoff washes away the dirt particles accumulated on the ledge. After hitting the ledge, runoff will drain away from the ledge either at its ends or directly over its front face. This will result in long stain streaks originating from the end of the ledge and/or short stain marks on the vertical face of the ledge.

An experimental simulation was conducted to investigate rain runoff flow. It has been determined that runoff flow off the length of a ledge would be even unless there are dents, joints or cracks along the ledge. Flow at the two ends of the ledge is more concentrated due to increased flow from runoff along the returns of the recessed unit. The experiment has reinforced the findings that these are areas where stains will be expected to manifest (Fig. 4.5). Runoff flow pattern over ledges therefore directly affects the staining pattern that will form around the ledge.

Stain marks on the face of the ledge will be evenly formed unless cracks, jointing or inclinations are found on the ledge to divert and concentrate runoff flow as shown in Figs. 4.6 to 4.13.

Figure 4.5. Runoff flow pattern off a ledge corresponds to likely areas where stains will form.

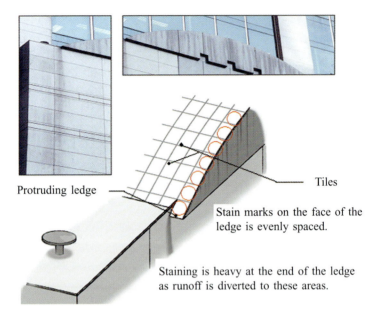

Protruding ledge

Tiles

Stain marks on the face of the ledge is evenly spaced.

Staining is heavy at the end of the ledge as runoff is diverted to these areas.

Figure 4.6. Staining originating from the end of the ledge.

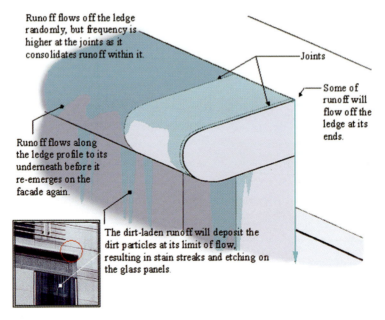

Runoff flows off the ledge randomly, but frequency is higher at the joints as it consolidates runoff within it.

Joints

Some of runoff will flow off the ledge at its ends.

Runoff flows along the ledge profile to its underneath before it re-emerges on the facade again.

The dirt-laden runoff will deposit the dirt particles at its limit of flow, resulting in stain streaks and etching on the glass panels.

Figure 4.7. Staining mechanism at the end of a ledge.

Figure 4.8. When access doors, ventilation holes and window openings are designed into facades, they must be incorporated properly so that any ledges that may be created will not be problem areas where dirt will accumulate and stains will originate.

4.2.3. Design Considerations

Logically, the easiest solution to prevent or minimise staining by ledges is to design facades without ledges or to put ledges on facades that receive little impacting rain. However, these would be too restrictive on designs.

A possible solution that is feasible would be to increase the outward inclination of the ledge to more than 30° to the vertical. This would increase the rate of flow of runoff when it flows over the ledge onto the wall below. The runoff would then have a stronger washing effect.

The ledge could also be made to slope inwards so that runoff does not flow over the ledge but is retained at the ledge. The retained runoff can then be channelled into a drainage downpipe constructed within the facade system and be drained off in a controlled manner (Fig. 4.14).

The ledge could also be made to extend further beyond the general facade by way of an attachment so that runoff is "thrown away" from the facade. Together with a throating at the underside of the extended ledge, the runoff can be prevented from trickling back to the facade, hence minimising staining (Fig. 4.15).

(a) Sloping ledges at the window promotes staining along the horizontal edges.

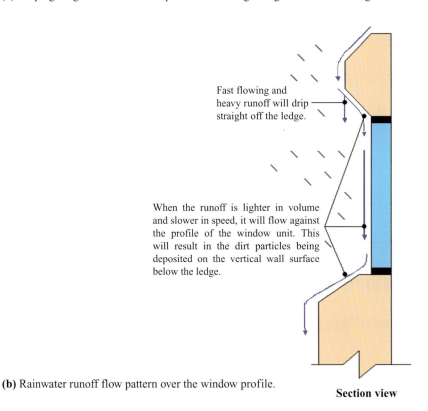

Fast flowing and heavy runoff will drip straight off the ledge.

When the runoff is lighter in volume and slower in speed, it will flow against the profile of the window unit. This will result in the dirt particles being deposited on the vertical wall surface below the ledge.

(b) Rainwater runoff flow pattern over the window profile.

Section view

Figure 4.9. Mechanism of staining along sloping window ledges.

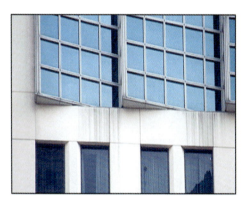

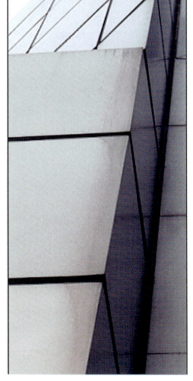

Figure 4.10. Runoff flowing from the top of the building tends to be concentrated along the sharp corners of the building due to wind effects. When the runoff continues its downward flow onto a protruding ledge, its rate of flow will be slowed down and the dirt particles will be deposited as stain marks.

Figure 4.11. Improper drainage from the roof has caused the roof to act as a ledge, allowing runoff to overflow and carry along any dirt retained on the roof to the facade.

Figure 4.12. Stain marks can be observed where there are expansion joints. The expansion joints act as channels and divert flow to it.

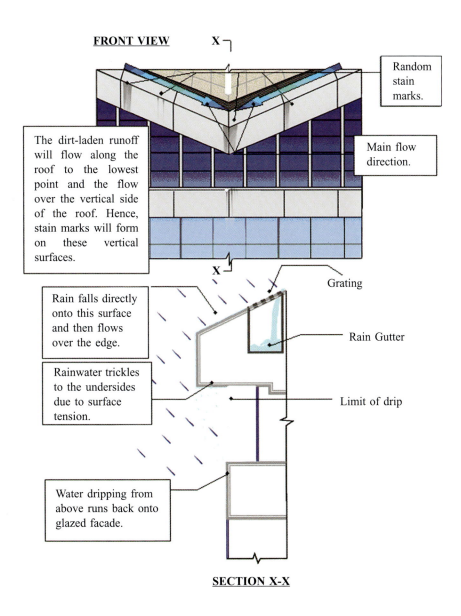

FRONT VIEW

Random stain marks.

The dirt-laden runoff will flow along the roof to the lowest point and the flow over the vertical side of the roof. Hence, stain marks will form on these vertical surfaces.

Main flow direction.

Grating

Rain falls directly onto this surface and then flows over the edge.

Rain Gutter

Rainwater trickles to the undersides due to surface tension.

Limit of drip

Water dripping from above runs back onto glazed facade.

SECTION X-X

Figure 4.13. Flow path of runoff after it falls off the roof ledge.

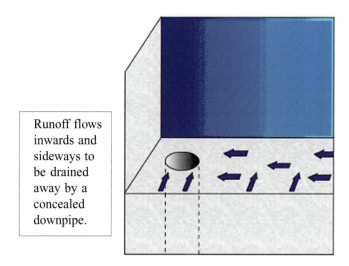

Runoff flows inwards and sideways to be drained away by a concealed downpipe.

Figure 4.14. Runoff gathered from the recessed window unit will flow inwards to the drainage down pipe.

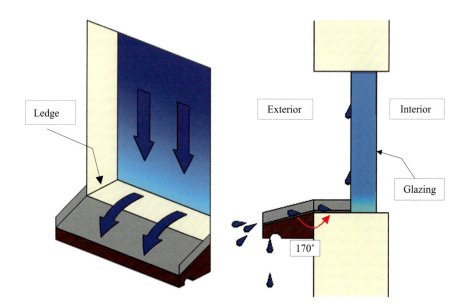

Figure 4.15. Ledge attachment to throw water off the ledge of a recessed window.

When the recession is rounded to form a ledge that concentrates runoff at a point, a joint or drainage track installed at the lowest point of the recession would be able to concentrate and direct runoff off the ledge and thus prevent staining (Fig. 4.16). Such design features may be useful on ledges where runoff always flows off at a particular point.

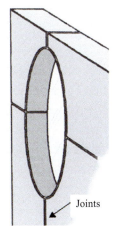
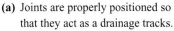
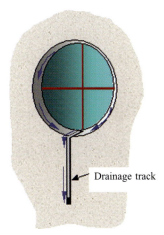

(a) Joints are properly positioned so that they act as a drainage tracks.

(b) Drainage track to channel runoff properly.

Figure 4.16. Preventing staining on rounded ledges where runoff will concentrate at the lowest point.

4.3 Joints Misalignment

In the construction of a facade, joints are essential to allow for either movement of panels or to provide for drainage of runoff on the facade [7]. Panel joints can either be open or infilled with an elastic material such as sealant. Regardless of their function, joints affect runoff flow and stains will form at areas where a file of vertical joints is broken. Other than between panel joints, joint misalignment can also occur between window joints or access panel door joints with joints on the general facade.

When facade components such as window units and access panels are of different sizes as the facade panels, their joints cannot be made to be in alignment, and a straight downward path of flow cannot be provided for the runoff . This prevents runoff from flowing in an undisturbed manner off the facade.

4.3.1. Factors Causing Staining at Misaligned Joints

Misalignment of joints has been observed on many buildings in Singapore and staining almost always occur on such design features.

Joints that run vertically or horizontally on the facade can be presented as irregularities on the facade and behave as crevices that trap and retain dirt particles. Under the combined effects of gravity and wind action, runoff will concentrate at these places of irregularities [7, 8, 16]. Wind causes a lateral migration of runoff, which concentrates the downward flow along the lines of vertical protrusions and depressions (Fig. 4.17) [11, 17]. Such areas are also points where dirt concentration is highest and would be carried away by runoff.

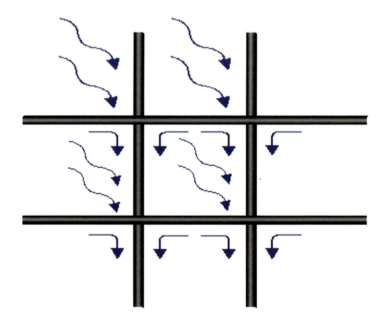

Figure 4.17. Runoff flow concentrated along irregularities such as joints. [11]

(a) Joints that prevent runoff from overflowing.

(b) Joints where runoff will overflow easily to result in staining.

Figure 4.18. Joints between panels.

When runoff concentrates to flow within facade joints, the construction of the joints will determine whether the runoff will be totally contained and directed to flow within the joints or overflow the joints to cause staining at particular areas. Joints that have the ability to effectively contain runoff so that it flows into and along it without overflowing the joints will cause little staining to occur as shown in Fig. 4.18(a). However, joints that channel runoff to flow into it easily but at the same time not able to retain runoff within it will encourage stains to form as shown in Fig. 4.18(b).

4.3.2. Mechanisms of Staining at Misaligned Joints

Findings from the rain simulation experiment have shown that when water droplets first impact on a dry panel, the resultant runoff flow is undeterminable and random. As the speed and volume increases, the separate streams concentrate into larger streams near to any vertical obstruction such as a vertical window or panel joints and flow along its inner sides [11]. When the vertical joint breaks at the next panel, runoff continues to flow downwards with a small volume being diverted sideways along the horizontal joint. This spilling over of runoff from the vertical joints onto the facade causes the rate of flow and volume to be reduced such that the runoff will no longer have the capacity and speed to carry the dirt particles along. Long stain marks will then form directly below the vertical joint and be bounded at both sides by shorter and less distinct stains (Fig. 4.19). When spraying lightens and stops, runoff flow becomes more distinct and corresponded to the existing stain marks. The limit of flow shortens as runoff dries up, allowing the dirt to build up onto the existing stain marks.

The sequence of staining as illustrated in Fig. 4.20 demonstrates that due to the mechanism of staining at misaligned joints, stain marks will be most intense at the area immediately below where the vertical joint discontinues and will decrease in intensity further away from the discontinuity.

Figures 4.21 to 4.25 show facades that experience staining due to misaligned joints.

For aesthetic reasons, expansion joints are sometimes inclined or misaligned to break the monotony of the facade. However, the inadequacy of the joints in containing and channelling runoff off the facade causes runoff to overflow the inclined or misaligned joints and continue vertically downwards to result in vertical stain streaks on facade panels and even glass. The vertical stain streaks will usually originate from an intersection point of the joints as shown in Figs. 4.24 to 4.26. Such points act as obstruction for the flow of runoff within the joint and cause the runoff to be "misplaced" and to overflow the joint.

4.3.3. Design Considerations

Expansion joints that are deliberately designed to be out of alignment may cause uncontrolled runoff flow. Streams of runoff flowing along the sides of a vertical joint would deposit the dirt it is carrying when it spills onto the panel below. Both closed and open expansion joints should be made to be in alignment throughout the height of the facade. Runoff flow

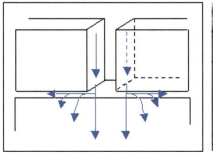 

(a) Diagrammatic flow of runoff within joints.

(b) Runoff runs directly over horizontal joints.

Figure 4.19. Flow of water in vertical and horizontal joints.

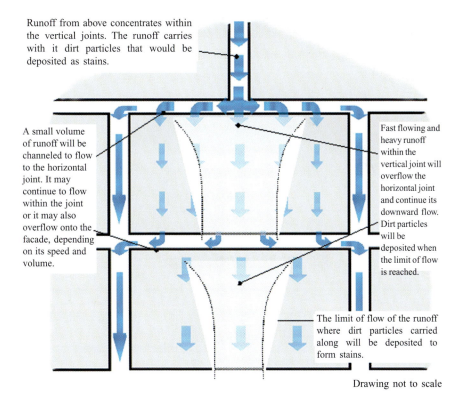

Runoff from above concentrates within the vertical joints. The runoff carries with it dirt particles that would be deposited as stains.

A small volume of runoff will be channeled to flow to the horizontal joint. It may continue to flow within the joint or it may also overflow onto the facade, depending on its speed and volume.

Fast flowing and heavy runoff within the vertical joint will overflow the horizontal joint and continue its downward flow. Dirt particles will be deposited when the limit of flow is reached.

The limit of flow of the runoff where dirt particles carried along will be deposited to form stains.

Drawing not to scale

Figure 4.20. Runoff flow pattern within typical facade joints.

Figure 4.21. Misalignment of panel joints resulting in stain marks.

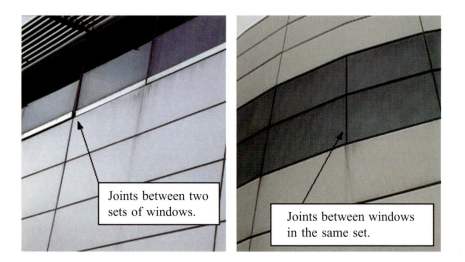

Figure 4.22. Joint misalignment between window joints and panel joints will result in stain streaks forming on the panels. Stains marks are absent where window sets meet because there is no discontinuation of vertical joints.

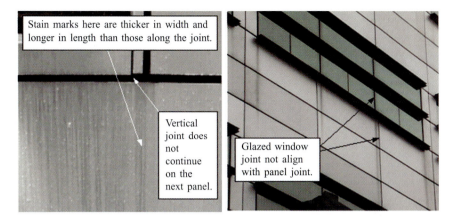

Figure 4.23. Abrupt discontinuation of joints result in stain marks.

Figure 4.24. Vertical joint does not continue the length of the building, thus causing staining where the joint breaks.

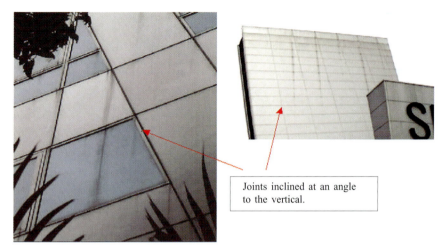

Joints inclined at an angle
to the vertical.

Figure 4.25. Joints inclined at an angle to the vertical.

(a)

(b)

Figure 4.26. (a) Joints misalignment on the facade of the building interrupts the even flow
of runoff and results in staining. (b) Drawing shows the clear misalignment of joints.

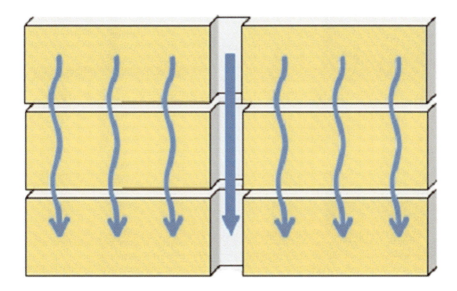

Figure 4.27. Runoff should be allowed to flow unobstructed along vertical joints.

can then be controlled to flow from the top of the facade to the bottom, and subsequently be drained off the facade as shown in Fig. 4.27.

Attention must also be paid to workmanship during the construction of these joints to ensure that they are in line. Poor workmanship can sometimes result in joints being misaligned, causing flow to be affected and bringing about long streaks of stains beneath where the misalignment occurs (Fig. 4.28). This may happen on tiled facades where on-site installation is heavy and misalignment resulting from poor workmanship may occur. Encouraging prefabrication and using facade systems that require little on-site installation may rectify such a problem [11].

4.4 Protruding fixtures

Any installation of signboards, logos, letterings, external lightings, speakers and other small fixtures onto the facade, done independently from the facade installation can be grouped as protruding fixtures.

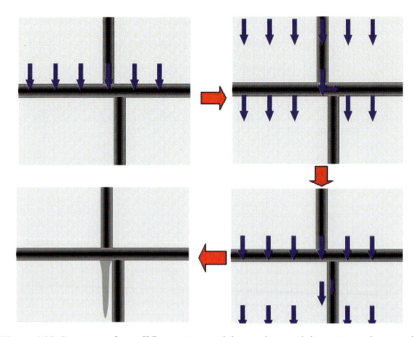

Figure 4.28. Sequence of runoff flow pattern and the resultant staining pattern when panels are slightly out of alignment.

Many building owners give identity to their buildings by installing their company logos or names on the front facade of the building. However such minor projections from the facade could disrupt the flow of water and redistribute it, washing off dirt at some areas and depositing dirt at others, causing an unsightly effect [1].

4.4.1. Factors Causing Staining at Protruding Fixtures

Designing and installing fixtures on a facade without considering the effects that they will have on the facade may yield adverse results (Fig. 4.29). When these fixtures are installed to protrude on a facade, dirt will settle on its horizontal surfaces. The degree of stain formation due to protruding fixtures will depend on the width of protrusion of the fixture and the presence of corners created between the fixture and the facade.

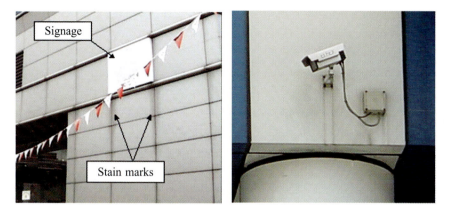

Figure 4.29. Protruding fixtures on building facade.

The more the fixture protrudes from the facade, the larger the horizontal surface area for dirt to settle. When runoff flows over the protrusion, the dirt will be washed off from the fixture and redeposited on the facade as stains. The protruding fixture causes staining in a very similar way as that of a ledge.

The method by which the fixture is attached to the facade has a large effect on the intensity of staining. Fixtures such as lightings and signage can be attached to the facade with clearance in between the two or be fixed directly to the facade without clearance. In the former, disruption of even and regular runoff flow on the facade will be minimised and dirt that gathers on the protrusion cannot be washed onto the facade. However, the clearance between the fixture and the facade must be wide enough to prevent debris and dirt particles from collecting in the clearance gap.

4.4.2. Mechanism of Staining at Protruding Fixtures

The basis of staining due to protruding fixtures can be reinforced with the observations obtained from subjecting a typical tightly installed light fixture to the rain runoff simulation experiment. The resulting staining pattern can be seen from Fig. 4.30 and similar staining patterns can be observed from Fig. 4.31.

The mechanism by which stain marks form around signage is very much similar to that of stains formed around protruding fixtures. This is illustrated in Fig. 4.32.

The horizontal and sloping surfaces present on the letterings are areas where more dirt can be retained to be later washed away to form stains on the facade. Stain streaks can originate either from the extreme vertical sides of the lettering or from behind the lettering. This is because of the different methods of attachment. The former case occurs when the

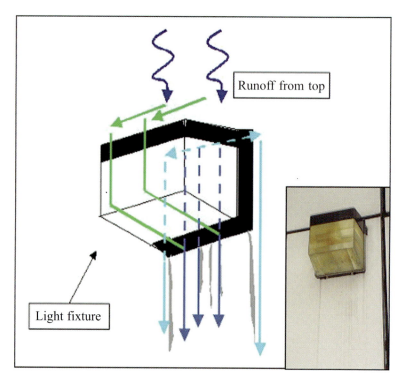

Figure 4.30. Mechanism of staining around a protruding fixture. (1) Runoff drips down vertically along the sides of the fixture. Long streaks of stains will be formed at both sides of the protrusion ("Moustache" staining) (Flow represented by →). (2) A few drips of runoff may flow behind the fixture and emerge below it if the attachment is not completely tight. This will cause a few long streaks to form at the area immediately below the protrusion (Flow represented by →). (3) Most of the runoff will trickle along the profile of the fixture, either dripping off its face or continuing to trickle back to the facade (Flow represented by →).

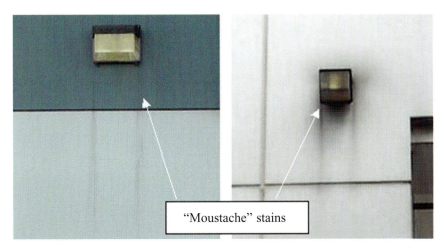

Figure 4.31. "Moustache" staining originating at the light fixture.

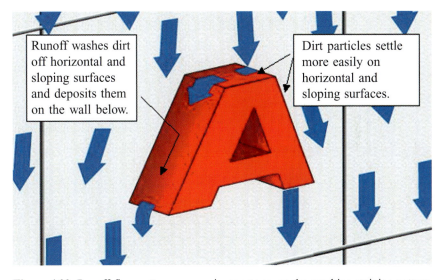

Runoff washes dirt off horizontal and sloping surfaces and deposits them on the wall below.

Dirt particles settle more easily on horizontal and sloping surfaces.

Figure 4.32. Runoff flow pattern over a signage to cause the resulting staining pattern.

letterings are fixed directly to the facade without clearance between the facade and the letterings, thus, dirt gathered on the lettering is washed onto the facade (Fig. 4.32). In the latter case, the letterings are not in direct contact with the facade and stains begin at the attachment points

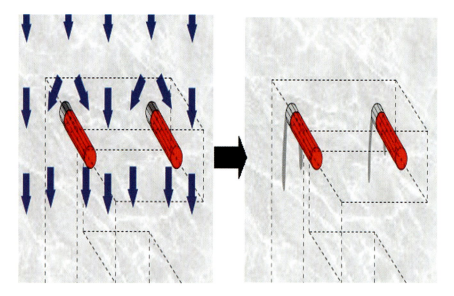

Figure 4.33. Staining originating from attachment points.

with no dirt from the lettering being washed onto the facade (Fig. 4.33). Staining in the latter case would thus be less intense as shown in Fig. 4.34.

4.4.3. Design Considerations

Protruding fixtures such as signage and external electrical installations are common protrusions that may cause staining to facades. Instead of fixing the letterings and signage tightly to the wall, they may be fixed by means of small-dimension bolts with a clearance between the facade and the fixture (Fig. 4.35). In this way, disruption of runoff flow would be significantly minimised. The attachment points would also gather minimal dirt.

By comparing the two methods of attaching fixtures to the facade as shown in Figs. 4.32 and 4.35, it is evident that by allowing a clearance between the facade and the fixture, runoff would be able to flow with minimal disturbance and staining below the protrusion would be minimised. When selecting the material for the attachment, more

Figure 4.34. The intensity of staining is much reduced when there is a clearance between the signage and the facade.

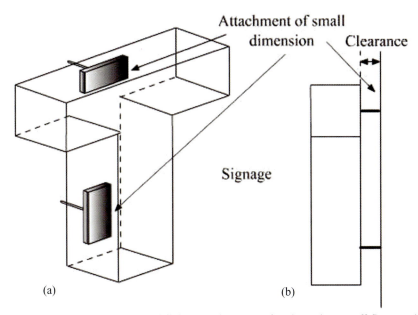

Figure 4.35. Proposed method of fixing attachments to facade so that runoff flow would not be disrupted. (a) Elevation view of signage been attached to the facade. Attachment used must be of small dimension. (b) Section view of signage being attached to the wall.

attention should be paid to its durability. Some contractor teams may overlook such considerations. Consequently, staining may be a result of the attachment material corroding, rather than due to dirt accumulation and redeposition [9–11].

4.5 Louver Units

Louver units provide visual texture to a building facade and at the same time serve as a functional component to provide ventilation or to expel air from the building. Needed mostly for circulation purposes, they have exacted a price in terms of visual deterioration and routine maintenance costs. They can be in the form of louver doors or windows. Louver units may also be designed as part of the facade solely for aesthetic purposes. The exhausted air from a louver unit usually contains a high concentration of dirt that is retained and accumulated at the top face of each louver pane. When runoff flows over the louver panes, the dirt particles are carried along and redeposited as stain marks. For this reason, areas below louver units are often stained. The staining effects of louver units are thus an important design consideration when designing facades.

4.5.1. Factors Causing Staining around Louver Units

The intensity of staining derived from louver units on the facade is dependant on a few factors:

- Surface area and angle of inclination of louver panes,
- Location of building and orientation of facade,
- Location of louver unit on the facade,
- Functional purpose of louver unit.
- Effective facade area where runoff can flow over after running over the louver unit.

The louver panes within a louver unit constitute a large surface area for dirt to settle. Since louver units are mostly used for ventilation

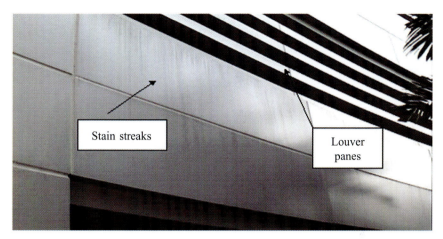

Figure 4.36. Louver units with horizontal louver panes.

purposes, it is expected that dirt-laden air will be forced out between the louver panes. The larger the surface area of panes, the greater the amount of dirt particles that will be retained. The angle of inclination of the panes would also affect the rate of settlement of dirt. Panes that are horizontal (Fig. 4.36) will be less prone to retaining dirt as compared to inclined panes since a portion of the inclined panes are shielded from the direct outflow of air (Figs. 4.37 and 4.38).

The rate at which dirt settles on the louver panes is higher than normal settlement on the other areas of the facade. Thus, louver units that are sited away from impacting rain or are recessed into the facade such that runoff or driving rain will not be able to reach the dirt retained on the louver will not cause staining. This allows dirt to build up evenly on the louver panes without disturbance from runoff flow.

Louver units may be used on a facade to break its monotony. It could also be utilised to ventilate spaces within the building. When used for ventilation purposes, air such as that from air-conditioning exhaust ducts will contain a higher concentration of dirt particles and thus the rate of accumulation of dirt on the panes would be higher.

When the louver unit is located on the ground floor or when it is located such that the runoff after flowing over the louver panes will drain off the facade immediately, staining may also be minimised.

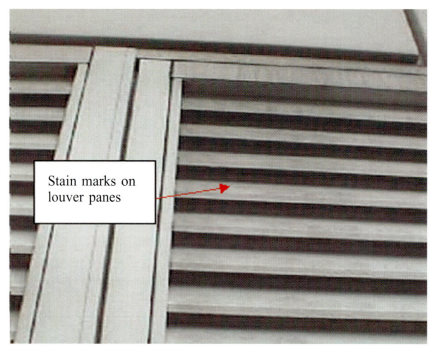

Figure 4.37. Stain marks on the louver panes of a louver door unit.

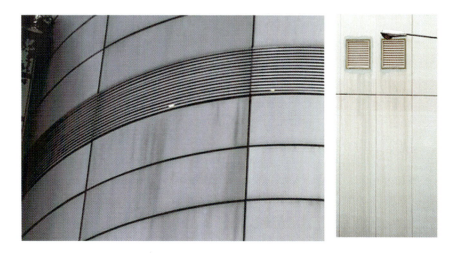

Figure 4.38. Serious staining on the facade due to louver units.

4.5.2. Mechanism of Staining at Louver Units

Staining at louver units occur due to runoff or driving rain carrying away and re-depositing the dirt particles that settle on louver panes. The sequence of staining is stepwise and begins with runoff flowing from facade areas above the louver unit onto the louver panes (Fig. 4.38). Driving rain may also impact on the louver panes to form runoff that carries off the dirt particles. From the topmost pane, runoff will continue to flow downwards to the last pane in no definite pattern of flow. This irregular flow causes washing off in some areas while promoting staining at others. When louver panes are inclined, only the outside portion of the panes experience wetting while the rest of the pane will be sheltered by the pane above it. The washing away of dirt from the outside portion of each louver pane results in the dirt being deposited on areas of the facade below the louver unit.

A rain simulation experiment conducted on a typical louver unit also demonstrated similar runoff flow pattern (Fig. 4.40). It was shown that the dripping of runoff down the louver unit was random and unpredictable and there was no defined location where drips were more concentrated.

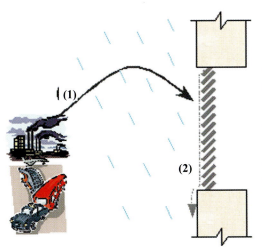

Louvers have a large effective horizontal surface area that can double as a dust fall collector from urban activities (1). The dust collected on louver panes would be washed off from the louvers and deposited as stains streaks on the immediate facade surface below (2).

Figure 4.39. Large louvers are able to collect more dust and result in stains more easily.

Figure 4.40. Random runoff flow pattern from louver pane to pane. Experimental setup.

However, there would be areas of flow concentration if any of the abovementioned design features (i.e. ledges, misaligned joints and protruding fixtures) were sited above the louver unit. These have the effect of disrupting the even flow of runoff so that flow becomes concentrated in streams before it runs over the louver panes. Runoff would then flow in defined flow paths over the panes to yield stain streaks at particular areas. The sequence of stain formation for a rounded louver unit is similar and is shown in Fig. 4.41.

4.5.3. Design Recommendations

Louvers are incorporated usually for the functional purpose of exhausting air in the building to outside the building. If they are necessary as part of the facade, they should ideally be located near to ground level or where there is no wide expanse of facade beneath it (Fig. 4.42). By doing so, there would not be sufficient area for the runoff to reach its limit of flow and thus dirt would not be deposited. This may involve re-evaluating the location of machine rooms and other service rooms.

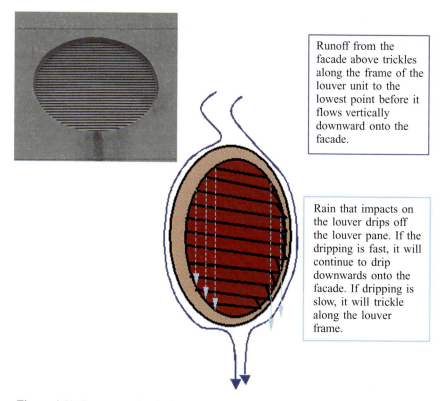

Runoff from the facade above trickles along the frame of the louver unit to the lowest point before it flows vertically downward onto the facade.

Rain that impacts on the louver drips off the louver pane. If the dripping is fast, it will continue to drip downwards onto the facade. If dripping is slow, it will trickle along the louver frame.

Figure 4.41. Sequence of stain formation at the lowest point of a round louver unit.

Louver units could also be recessed deep into the facade so that runoff flowing down from above will not run over the louvers and driving rain could not impact on the louver unit (Fig. 4.43). An up-stand can also be created at the horizontal recess so that run-off will not flow back onto the facade (Figure 4.44). Small louver units could be sheltered in this manner but for large louver units, the depth of recession may be too large and thus might not be feasible.

In the next section, an overview of the experimental setup designed to investigate the relationship between runoff flow and the pattern of staining would be given. The staining mechanism for ledges, joints misalignment, protruding fixtures and louver units will be investigated based on case studies and supported with experimental simulations.

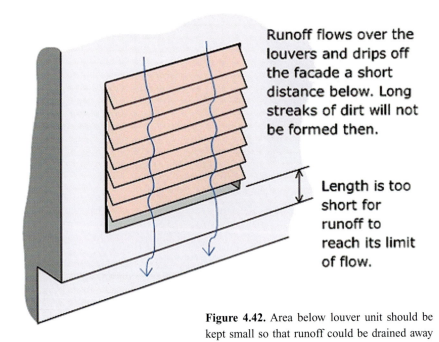

Runoff flows over the louvers and drips off the facade a short distance below. Long streaks of dirt will not be formed then.

Length is too short for runoff to reach its limit of flow.

Figure 4.42. Area below louver unit should be kept small so that runoff could be drained away before it reaches its limit of flow.

4.6 Experimental Methodology

To verify the pattern of staining around a design feature caused by runoff flowing over it, an experimental setup capable of reproducing impacting rainfall is devised as shown in Fig. 4.45. The setup was devised based on experimentally sound models used to study wind-driven rain patterns over a building facade [17–22]. The subsequent runoff flow pattern generated over a typical facade feature can then be investigated [23].

A nozzle was used to produce a full-cone spray pattern completely filled with spray droplets (Fig. 4.46 and 4.47). This nozzle is able to provide a round coverage. By using a combination of a few such nozzles, impacting rain can be realistically simulated with overlapping areas representing areas receiving heavier rainfall. The feature under test can thus be subjected to similar on-site conditions of irregular rain volume

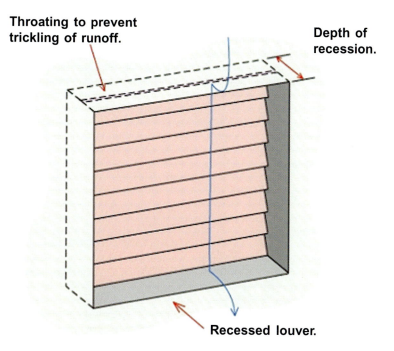

Throating to prevent trickling of runoff.

Depth of recession.

Recessed louver.

Figure 4.43. Runoff flow pattern after recessing the louver unit into the facade.

and flow rate. The subsequent flow pattern over the feature can thus be a combination of impacting sprays and runoff from above.

4.6.1. Methodology

With the experimental setup, facades that contain representations of the design features concerned were subjected to the rain simulation test. Observations were made on the runoff flow pattern 5 and 60 seconds after the water supply has been turned on, and 5 seconds and 15 minutes after the water supply has been turned off [23]:

- **5 seconds after water supply is turned on:** Synonymous with when the first few droplets of rain impact the facade and the initial runoff flows over a dry facade.

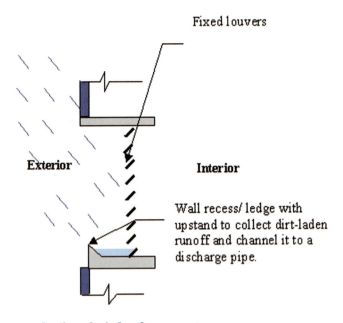

Fixed louvers

Exterior

Interior

Wall recess/ ledge with upstand to collect dirt-laden runoff and channel it to a discharge pipe.

Section of window louver system

(a)

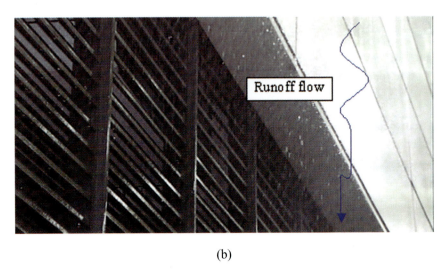

Runoff flow

(b)

Figure 4.44. (a) Up-stand at horizontal recess (cross sectional view). (b) Louvers to be recessed into the facade to minimise surface flow of water over them.

Figure 4.45. Experimental setup.

Figure 4.46. Nozzle used to simulate rainfall.

- **60 seconds after water supply is turned on:** Even wetting would have occurred and runoff flow would be regular and continuous. Synonymous with a heavy downpour at its heaviest intensity.
- **5 seconds after water supply is turned off:** Generation of runoff would have ceased; runoff flow would have slowed down and beginning to dry out. Synonymous with the period immediately after a light rain has stopped.
- **15 minutes after water supply is turned off:** The design feature would have nearly dried out and stain marks would have become more visible.

This simulation was used for the investigation of the various design features mentioned in the preceding sections.

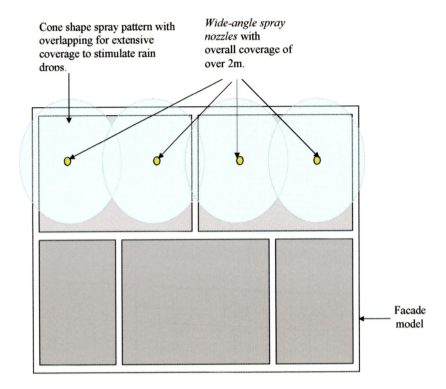

Figure 4.47. Spray pattern on facade using spray nozzles.

References

[1] P. Parnham, *Prevention of Premature Staining of New Buildings*, E. & F.N. Spon, London, 1997.

[2] W. H. Ransom, *Building Failures: Diagnostics & Avoidance*, E. & F.N. Spon, London, 1987.

[3] D. Campbell-Allen and H. Roper, *Concrete Structures: Materials, Maintenance and Repair*, John Wiley & Sons, New York, 1991.

[4] L. G. W. Verhoef, *Soiling and Cleaning of Building Facades*, Report of the Technical Committee 62 SCF, RILEM, Chapman and Hall, London, 1988.

[5] C. Briffett, "The performance of external wall system in tropical climates", *Energy and Buildings*, Netherlands, 1990.

[6] C. Briffett, "External finishes- Case studies on problems and solutions", Building Protection Conference, Proceedings of *Inter-Faculty Conference on Protection of Buildings from the External Environment, Paper 2*, 1987.

[7] C. Hall, "Absorption and shedding of rain by building surfaces", *Building & Environment* Vol. 17, pp. 257–262, 1982.

[8] R. Cooper, "Factors affecting the production of surface runoff from wind-driven rain", *RILEM International Symposium*, Rotterdam del. 1.1.2., 1974.

[9] E. B. Feldman, *Building Design for Maintainability*, McGraw-Hill, New York, 1975.

[10] N. G. Marsh, "The effect of design on maintenance", *Development in Building Maintenance- I*, Applied Science Publications Ltd., New York, 1979.

[11] M. Y. L. Chew, C. W. Wong and L. H. Kang, *Building Facades: A Guide to Common Defects in Tropical Climates*, World Scientific Publishing, Singapore, 1999.

[12] H. P. Teo and P. Ng, "External cladding defects in Singapore" in *Southeast Asia Building,* March, 1992.

[13] M. B. Ullah, "Analysis of rain and wind for building design", *National University of Singapore Seminar on Wind & Rain Penetration in Buildings,* 1994.

[14] M. C. Baker, "Rain deposit, water migration and dirt markings on buildings", *RILEM/ ASTM/ CIB Synposium on Evaluation of the Performance of External Vertical Surfaces of Buildings*, pp. 57–66, 1977.

[15] H. Ishikawa, "The extent of shelter provided by projections on external walls from driving rain", *RILEM/ ASTM/ CIB Symposium on Evaluation of the Performance of External Vertical Surfaces of Buildings,* pp. 154–169, 1977.

[16] O. Beijer, *Weathering on External Walls of Concrete,* Swedish Concrete Research Council, Swedish Cement and Concrete Research Institute, Stockholm. 1980.

[17] M. Y. L. Chew, "Efficient maintenance: Overcoming building defects and ensuring durability," *Conference on Building Safety*, The Asia Business Forum, Kuala Lumpur, 4 & 5 April 1994.

[18] W. H. Melbourne, "Cross-wind response of structures to wind action", Proceedings of *4th International Conference on Wind Effects on Buildings and Structures*, pp. 343–358, Heathrow, 1979.

[19] D. Inculet, D. Surry and P. F. Skerlj, "The experimental simulation of wind and rain effects on the building envelope", *International Conference on Building Envelope System Technology*, Singapore, 1994.

[20] E. C. C. Choi, "Simulation of wind-driven rain around a building", *Journal of Wind Engineering and Industrial Aerodynamics,* 46, pp. 721–729, 1993.

[21] G. Robinson and M. C. Baker, "Wind-driven rain and buildings", Technical Paper 445, National Research Council of Canada, Division of Building Research, Ottawa, 1975.

[22] M. Y. L. Chew, "A modified on-site water chamber tester for masonry walls", *Construction and Building Materials*, 15(7), 2001.

[23] M. Y. L. Chew and Tan P. P., "Facade staining arising from design features". *Journal of Construction and Building materials*, 17(3), 181–187, 2003.

CHAPTER 5

MAINTENANCE ASPECTS

5.1 General

The facade, being the essential architectural and functional component of a building, requires utmost attention on its long-term maintainability, reliability and operability. Any negligence considering the maintainability during the planning, design, procurement, construction, and start-up of facility can lead to increased life-cycle operation and maintenance costs, and expensive maintenance retrofits. Maintainability refers to the probability that a failed item will be restored to operational effectiveness within a given period of time, when the repair action is performed in accordance to the prescribed procedures [1]. It represents a goal-attainment evaluation that may assure better long-term performance of buildings throughout their life cycles in three aspects; end-user expectations, operation and maintenance [2–11].

Due to the fact that maintenance cost of a building during its service life constitutes a large proportion of the total costs, it is essential to give adequate considerations to the whole-life performance of a building. Table 5.1 shows an example of the factors to be considered at the design, construction and operational stages of a building facade in order to achieve maintainability.

5.2 Cleaning Stains

This section will briefly describe simple cleaning methods on stained facades.

Table 5.1. Maintainability of facade — Performance matrix.

COMPONENT : FACADE Component detail :	Integrity	Acoustics	Lighting	Thermal	IAQ	Visual
DESIGN AND CONSTRUCT						
1. Design						
a. Buildability (BDAS)						
b. Upgradability						
c. Retrofitability						
d. Accessibility						
2. Detailing						
a. Interface						
b. Joints						
c. Drainage						
d. Openings						
e. Penetration						
f. Fixtures						
3. Materials						
a. Durabilty						
b. Life cycle						
c. Repairability						
d. Replaceability						
e. Sustainability						
4. Construction						
a. Prefabrication						
b. Workmanship						
c. Warranty						
d. Commissioning						
OPERATIONAL						
1. Cleaning						
a. Material						
b. Labor						
c. Access						
d. Down time						
e. Loss in productivity						
f. Frequency						
2. Energy						
3. Inspection						
a. Frequency						
MAINTENANCE						
1. Repair						
a. Diagnostic						
b. Materials						
c. Labor						
d. Down time						
e. Loss in productivity						
f. Frequency						
2. Replace						
a. Displacement-neighboring components						
b. Materials						
c. Labor						
d. Down time						
e. Loss in productivity						
f. Frequency						

5.2.1. Exposed Brick and Concrete Walls

Bricks are usually durable materials that age gracefully and require minimal cleaning. Nevertheless, dirt staining and efflorescence do cause bricks to lose their aesthetic value over time [12]. Table 5.2 shows some cleaning methods for brick walls of various conditions [12].

Dirt stains and biological growth could be water blasted to remove them from the brick wall. Efflorescence has to be brushed or scraped off from the surface whenever the salts appear. Brick masonry can generally be cleaned with chemicals in conjunction with water rinsing. Acidic

Table 5.2. Cleaning and maintenance of brick wall [12].

Condition of Brick Wall	Cleaning Method
Dirt on brick	
Clay bricks	Organic stains should first be removed with detergents before other chemical agents are used. If the texture of the brick is rough, pressurised water cleaning (< 700 psi) may be used in conjunction with light brushing with fibre brushes, but taking care not to initiate efflorescence.
Glazed bricks	Surface soiling may be cleaned using a water rinsable neutral liquid detergent. Soiling beneath cannot be removed.
Biological staining	Removed by brushing with fibre brushes or in conjunction with water spray or chemical cleaning using a solution of muriatic acid.
Green/buff or cream coloured stains from Vanadium salts (newly erected brickwall)	Scrub with a solution of 10% hydrochloric acid containing detergent at 0.1% of the total acid solution and wash thoroughly with water. Leave the wall alkaline by washing with potassium hydroxide (50g/L).
Paint on wall	May be removable with water rinsable paint removers to BS 3761: 1995. Alkaline-based agents may also be useful. The surface should be rid of residual resins and pigments by washing with a high-pressure water lance.
Walls with efflorescence	Should be allowed to weather away over time. May be removed by dry brushing with a stiff bristle brush and rinsing with water. The residue should not be allowed to re-enter the brickwall at lower levels. Chemicals should not be used.
Walls with lime stains	Washing with dilute acid. The wall should be wetted with water before the acid is brushed on. Upon removal of the stains, the wall must be rinsed clean with water again.
Iron and manganese stains (light brown to black)	Chemical cleaning using 5% or 10% hydrochloric acid or painting the stains over with oxalic acid solution (120g/L).

cleaners containing dilute mineral acids such as hydrochloric, hydrofluoric, phosphoric and/or organic acids such as acetic and citric acids are used to remove heavy soiling from most brick masonry walls. The cleaning process involves applying the diluted cleaner to the pre-wetted surface and allowing a short dwell period. Chemical and soiling residues are removed by pressure water rinsing. The use of acid should be avoided in cases when steel accessories are incorporated in the brickwall. Bricks with high iron content may also be reactive to acids. Sometimes alkaline cleaners are used to remove soiling on brick surfaces, but the type of chemical cleaner should be selected with care, and then tested on small areas before it is used to clean the whole area. Certain chemicals may cause staining on some types of bricks. Soft bricks are particularly vulnerable to damage from aggressive cleaning methods. Cleaned bricks should be coated with a water repellent sealer to prevent bricks from getting wet [12–14].

If staining affects a large portion of a facade, it may be more economical to paint over the brick surfaces as shown in Figs. 5.1 and 5.2.

Exposed concrete is prone to staining due to surface irregularities [15]. Regular cleaning once or twice a year, with high-pressure water jet or a non-toxic and mild acidic-based solution is required to keep the building from staining [12–14]. Stains caused by rain that washes the dirt and dust, and algae growth are generally repairable by simple washing and scrubbing at the stained area (Table 5.3) [16, 17].

5.2.2. Natural Stone

Natural stones may be generally classified into two general categories based on its composition: siliceous or calcareous. The former tends to be more durable and relatively easy to clean with neutral cleaning solutions (Fig. 5.3). The latter is sensitive to acidic cleaning agents and requires cleaning with specially formulated stone cleaners.

Natural stones are basically porous and they absorb spills and stains if left untreated. Sealers with repellency properties against weathering and ultra-violet rays may be considered to minimise cleaning work.

Figure 5.1. Painting over a brick wall.

Figure 5.2. Painted brick wall.

Table 5.3. Cleaning and maintenance of concrete wall [16, 17].

Condition of Concrete Wall	Cleaning Method
Atmospheric soiling	Low-pressure water washing from top down. High pressure water jets should not be used as it may drive the stain further into the concrete. If insufficient, it could be supplemented by the following in order: brushing with a soft brush, a mild soap, a stronger soap, ammonia or vinegar.
Severe soiling	Chemical cleaning Ammonium hydroxide, sodium hypochlorite or hydrogen peroxide may be used with dilution. The surface should be flushed with water before and after washing to prevent etching by acidic agents. Chemicals containing salts may damage concrete due to adverse reactions. Mechanical cleaning Involves power tools such as grinders, buffers, chisels, brushes or steam/ flame cleaners. Concrete may be removed along with the stain to result in a roughened or uneven surface. Organic stains that cannot be removed with solvents may be burnt off with flame cleaners. However, the heat may cause part of the concrete surface to scale off.
Biological staining	Heavy growth should be removed by brushing with stiff fibre brushes, wooded spatulae, scrapers or a low-pressure water lance. Biocide should be applied to inhibit further growth.
Oils stains	May be removed by applying an emulsifying or degreasing agent. Deep stains should be poulticed with white spirit or trichloroethane. The deposits should be then removed with hot water pressure lance or with steam.
Walls with efflorescence/ lime stains	Usually disappears itself by natural weathering. May be removed by washing with a 5% solution of hydrochloric acid. Alternatively, brushing with soft compact brushes and sponging the residual powder away may be more effective since excessive wetting with water may initiate further efflorescence formation.

Figure 5.3. Cleaning of facade consisting of glass and natural stone.

Cleaning by water coupled with scrubbing or high-pressure water jet could effectively remove most of the stain from stone cladding surfaces (Table 5.4) [12]. Cleaning should begin at the top so that excess water can run down and pre-soften the dirt below. Acidic cleaning agents should not be used for granite as they may attack the pyrite (iron sulphide) which is inherent in granite to result in brown stains. It is also not proper to use cleaners that contain petroleum (which may change the appearance of the stone) or products that contain other acids or abrasives that may scratch the surface [18, 19].

5.2.3. Tiles

Table 5.5 summarises the common staining problems with external tiled wall and shows examples of maintenance strategies [20, 21]. The dirt stains will be more concentrated at the mortar joints since mortar has a higher porosity and absorbs water quickly to leave behind dirt particles within the joints. If a sealant is used at movement joints and the sealant

Table 5.4. Cleaning and maintenance of natural stone wall [12].

Condition of Natural Stone Wall	Recommended Cleaning Method
Unpolished Granite (Atmospheric Soiling)	Chemical/ abrasive methods needed. Agents containing hydrofluoric acid may be useful. Alternatively, use alkaline cleaners followed by neutralisation with weak organic acid.
Polished Granites (Atmospheric Soiling)	May be removed with non-ionic soap and scrubbing in water. Surfaces should be thoroughly rinsed and wiped dry to prevent water spotting. * Visual inspection every five years. If necessary cleaning, repointing and surface repairs in accordance with BS 6270:Part 1 * Strong acidic cleaning agents should not be used as it attacks pyrite (iron sulphide) inherent in granite to result in a brown stain. * Apply silane-based impregnating agent every 5 years to seal stone against dirt and pollutants.
Marble Water soluble sooty particles	Washing with small quantities of water. Soften the dirt by hand-spraying, followed by scrubbing with bristle brush and hand-spraying to remove dirt.
Metallic, oil or grease stains	Remove with liquid detergents. It stains persist, use acid or alkali based agents. * Acids, phosphorus, chlorine or scouring powder should not be used. Hard water will encourage discolouration, particularly if iron is present, and cause the build-up of insoluble salts. Re-polishing may be required on a regular basis.
Sandstone	Chemical cleaning using hydrofluoric acid and orthophosphoric acid-based agent. Or dry air abrasion cleaning using mineral slag abrasive agents.
Stones with Efflorescence/ Lime Stains	Remove by brushing with fibre brushes or in conjunction with water spray or chemical cleaning using a solution of muriatic acid.
Biological Staining	Remove by dry brushing with wooden scrapers or bristle brush or by high pressure water jetting. Surface should then be treated with anti-fungicidal wash.

* denotes general notes

Table 5.5. Cleaning and maintenance of tiled wall [20, 21].

Condition of Tiled Wall	Cleaning Method
Ceramic Tiles	General cleaning by wiping with wet cloth or scrubbing with sponge. For heavier soiling, use a mild detergent solution and leave it on the surface for 5 minutes before scrubbing lightly with a brush.
Mosaic Tiles	General cleaning by wiping with damp sponge mop. For heavier stains, cleaning agents can be supplemented with brushing. Pressure blasting can also be considered to wash away dirt trapped at the joints.
Efflorescence/ Lime Stains	Removed by dry brushing or with water and a stiff brush. Heavy efflorescence or lime stains may be removed with mineral acids such as hydrochloric, sulphuric and nitric or other organic acids. Wet the surface well before and after the solution is applied.
Mildew	Remove with a dilute solution of ammonia or bleach. Concurrent scrubbing may be needed.
Biological Staining	Use a weak acid such as vinegar.

fails and becomes sticky, it will hold dirt until it rains. Dirt is then deposited in streaks down the building, emanating from that point [19]. Depending on the tiles used and the extent of staining on the surface, cleaning agents can be selected according to the state of staining. Care should be taken during the selection of cleaning agents as abrasive agents can easily etch the tiles, making them more vulnerable to dirt. Tiles used should be glazed and of darker colour to mask excessive dirt stains.

5.2.4. Metal

The appropriate type of cleaning method used is determined by the degree of soiling, the size, shape and location of the surface to be cleaned. The cleaning specifications should be followed closely with respect to the frequency and method of cleaning (Table 5.6) [22, 23].

It is preferable to clean metal surfaces in the shade as possible chemical reactions on hot metal surfaces may be highly accelerated and

Table 5.6. Cleaning and maintenance of metal wall [22, 23].

Condition of Metal Wall	Cleaning Method
Aluminum	
Anodic Coating	
Lightly soiled	Flush surface with water at moderate pressure. Use mild detergent, and brushing or sponging concurrently if necessary.
Heavy soiling	Scrub with a nylon-cleaning pad wet with surface protective material. Rinse surface with water and wipe dry with a chamois, squeegee or lint-free cloth or air dry.
	Power cleaning tools (e.g. air-driven reciprocating machine fitted with abrasive pads) and mild detergent can also be used. Direction of travel of machine with respect to geometric configuration of the surface being cleaned should be noted.
Powder Coating	
Lightly soiled	Flush surface with water at moderate pressure. Use mild detergent, and brushing or sponging concurrently if necessary.
Heavy soiling	Mild solvent (eg. mineral spirits) may remove grease, sealant or caulking compounds.
	Dried concrete stains can be removed with diluted muriatic acid (under 10%). Vigorous rubbing with non-abrasive brushes or plastic scrapers may be necessary
	* Some solvents may extract materials from sealants and cause staining or damage the sealant.
	* Abrasive cleaners containing ketones, ethers or alcohols and steel wool should not be used.
	* Coatings may require re-decoration after 10–15 years.
	* Clean polluted/marine areas every 3 months, other areas every 6 months.
Steel	
Dirt and pollution	May be removed by rinsing with water and mild detergent, then scrubbing in the direction of the grain with soft cloths, sponges, fibrous brushes, or plastic pads. Abrasive actions should however be minimised since these may scratch the finish. Steel wool/brushes will causing pitting and should not be used.
Chlorides	Remove with warm water.
Fingerprints, grease, oil	Remove with a combination of water, mild detergent, and mineral spirits.

Table 5.6. (*Continued*)

Condition of Metal Wall	Cleaning Method
Iron stains (from bolts, screws, etc)	Remove rusty elements and immerse them in nitric acid.
Graffiti (Water-soluble inks)	Use warm water and a non-ionic detergent and rinsing with water.
Other inks and paint	Use a combination of water, non-ionic detergent and mineral spirits and rinsing with clean water.
Lead pencil markings	Remove with an oily cleaner such as paste wax.
Weld stains	Remove with mild abrasive cleaner in paste form and water.
Corrosion products	Remove with warm water, detergent, and plastic pads. If severe, mechanical methods (e.g. grinding or sandblasting) may be appropriate. Surface refinishing with fine abrasives to restore to original.
	* Clean polluted/marine areas every 3 months, other areas every 6 months.
	* Annual inspection and maintenance in accordance with BS 5427: Part 1, Table 9.
Biological Staining	Fungicide. Leave on wall for up to 7 days. All traces of the fungicide and effluent should be removed and the surfaces thoroughly rinsed with water.

*denotes general notes

non-uniformity can occur. For slightly soiled surfaces, cleaning should be done with water and some detergent. It should be completed by carefully rinsing with clear water and wiping with a soft and absorbent cloth. For moderately soiled surfaces where the soiling cannot be removed by normal detergents, it is recommended that products which are developed especially for this purpose be used. These products may contain detergents and very light abrasive materials. For very dirty surfaces where the dirt is very stubborn, it may be necessary to apply the same means as mentioned above but with the use of synthetic pads.

There are many ways to clean metals, from using plain water to harsh abrasives. The mildest possible method should be used, particularly for anodised aluminum. The following cleaning materials and procedures are listed in ascending order of harshness. The mildest treatment should be tried on a small area and if the results are not satisfactory, the next method may be examined.

- Plain water.
- Mild soap or detergent.
- Solvent cleaner, e.g. kerosene, turpentine, white spirit.
- Non-etching chemical cleaner.
- Wax-based polish cleaner.
- Abrasive wax.
- Abrasive cleaner.

The procedure for cleaning should begin with applying the cleaning solution only to an area that can be conveniently cleaned without changing position. The surface should be thoroughly rinsed with clean water before applying the cleaner. Cleaner rundown should be minimised on the lower portions of the building and such areas should be rinsed as soon and as long as it is practical. The metal panel should be dried with a clean cloth to prevent streaking. There should be no concentration of the cleaner at the bottom edges of the aluminium. If abrasives are used then the appearance of the metal finish may be altered. If there is a grain in the finish then cleaning should always be with the grain. When all other methods fail it may be necessary to resort to heavy-duty cleaning. This involves the use of cleaners containing strong etching chemicals or coarse abrasives (Fig. 5.4) [22–24].

Cleaners containing strong organic solvents will have a deleterious effect on organic overlay coatings, but not on anodised aluminium. However, the possibility of solvents extracting stain-producing chemicals from sealants and affecting the function of the sealants must also be considered [19].

Depending on the causes of staining, different measures can be taken to mitigate its influence. In most cases, only cleaning work is needed. However, if the staining is caused by the degradation of cladding materials and sealant, the replacement of the defective element is needed.

Figure 5.4. Cleaning aluminum panels with a scrub.

5.2.5. *Glass*

Table 5.7 shows the common maintenance methods for glass facades [25, 26]. The procedure for cleaning glazed surfaces begins by wringing a cloth, sponge, or chamois until it is almost dry before wiping the glass surface (Fig. 5.5). The wet surface is then dried with newspapers, paper

Table 5.7. Cleaning and maintenance of glazed surfaces [25, 26].

Condition of Glass Wall	Cleaning Method
Soiled and greased glass	Wring out a cloth, sponge, or chamois almost dry before wiping the glass surface. Use an alkali, such as ammonia or baking soda or washing soda. Dry the wet surface with newspapers, paper towels, window wipes, or a chamois. Avoid washing windows in direct sunlight because they tend to streak and are more difficult to clean.
Hard water deposits and soil	Use a weak acid such as vinegar (a strong acid would etch the glass). Avoid cleaning glass under direct sunlight as they tend to streak.

Figure 5.5. Wiping glass surface with chamois.

towels, window wipes, or a chamois. Avoid washing windows in direct sunlight because they tend to streak and are more difficult to clean.

5.2.6. Plaster-and-Paint Surfaces

Regular inspection of the coated surfaces is important in determining the seriousness of stains and the frequency of cleaning required.

The use of alticides, washing and repainting of walls regularly (every few years) can keep stains and biological growth at bay. Care must be taken on painted surfaces as sunlight can actually penetrate through one or two layers of paint and cause growth underneath instead, and eventually break through the new layer of paint [27].

Stained plaster-and-paint walls are usually cleaned by washing and scrubbing (Table 5.8) [28, 29]. However if the stains are too serious and widespread, it is more appropriate to remove the affected surface coating, sand, clean and redo the coating.

Table 5.8. Cleaning and maintenance of painted wall [28, 29].

Condition of Painted Wall	Cleaning Method
Oil paint	Wash with water and a non-abrasive weak alkaline detergent (e.g. hand dishwashing liquid). Stronger solutions may remove some of the paint.
	Walls should be cleaned upwards commencing from the bottom and in a sideways manner. This prevents staining of the surface by water.
Semi-gloss paints	Wash with a non-abrasive weak alkaline detergent using less water. Hard-bristled brushes should not be used.
	Walls should be dry brushed regularly and washed only once a year.
Gloss paints	Water and a weak detergent may remove atmospheric dirt on the surface. Rinse with clean water after washing.
Water paints	Small build-up of dirt may be removed using light sponging with a weak detergent and clean water. Excessive water should not be used as it may damage the finish. Regular cleaning by dry brushing is recommended.

5.3 Facade Access Systems

A facade access system is required for inspection, maintenance and cleaning of tall buildings. Such systems can be expensive to provide and maintain, difficult to design, dangerous to use and awkward to operate especially for buildings with high irregularity in shape (Fig. 5.6). The facade access system shall be designed, fabricated and installed to satisfy the following minimum requirements [11, 30, 31]:

- It should have a relatively long design life of between 10 to 25 years.
- It should withstand all loadings (wind, dead and live) as stipulated by local codes and specifications.
- It should provide workers with easy and safe access to all areas of the facade for all cleaning, repair and replacement works.
- When in operation, it should not bring about disruptions to tenants' activities or cause damage to the facade.

Figure 5.6. A specially designed gondola to reach inaccessible areas.

- It should be designed to provide maximum coverage to the facade including clearing protrusions on the facade such as sunshades and hence fulfil the required cleaning cycle.

The provision of facade access systems should be carried out in strict accordance with recognised codes and standards.

5.4 Codes and Standards

Relevant codes and standards pertaining to the use of equipment and cleaning of the external facade that ensure the safety and health of maintenance personnel are:

International Codes and Standards:

➢ **BS 7883:1997** Code of practice for application and use of anchor devices conforming to BS EN 795.

➢ **BS EN 795:1997** Protection against falls from a height. Anchor devices. Requirements and testing.

➢ **BS 5973:1993** Code of practice for access and working scaffolds and special scaffold structures in steel.

➢ **BS 6037:1990** Code of practice for permanently installed suspended access equipment.

➢ **BS 1139-1.2:1990** Metal scaffolding. Tubes. Specification for aluminium tube.

➢ **BS 1139-2.1:1991, EN 74:1988** Metal scaffolding. Couplers. Specification for steel couplers, loose spigots and base-plates for use in working scaffolds and falsework made of steel tubes.

➢ **BS 1139-2.2:1991** Metal scaffolding. Couplers. Specification for steel and aluminium couplers, fittings and accessories for use in tubular scaffolding.

➢ **BS 1139-3:1994** Metal scaffolding. Specification for prefabricated mobile access and working towers.

➢ **BS 1139-4:1982** Metal scaffolding. Specification for prefabricated steel splitheads and trestles.

➢ **BS 1139-5:1990, HD 1000:1988** Metal scaffolding. Specification for materials, dimensions, design loads and safety requirements for service and working scaffolds made of prefabricated elements.

➢ **BS 6399-2:1997** Loading for buildings. Code of practice for wind loads.

➢ **BS EN 1808:1999** Safety requirements on suspended access equipment. Design calculations, stability criteria, and construction. Tests.

➢ **BS EN 280:2001** Mobile elevating work platforms. Design calculations. Stability criteria. Construction. Safety. Examinations and tests.

➢ **AS/NZS 4488.1:1999:** Industrial rope access systems — Specifications.

➢ **AS/NZS 4488.2:1997:** Industrial rope access systems — Selection, use and maintenance.

Singapore Codes and Standards:

> SS 210: 1979 Industrial safety belts and harnesses.
> SS CP 14: 1996 Code of practice for scaffolds.
> SS CP 20: 1999 Code of practice for suspended scaffolds.
> SS 280: 1984 Frame scaffoldings.
> SS 311: 1994 Steel tubes and fittings used in tubular scaffolding.
> The Factories (Building Operations and Works of Engineering Construction) Regulations 1998.

5.5 Means of Access

A building facade's maintenance expenditure is influenced by the building's shape and form. The form of buildings that have been chosen may have a bearing on the method of maintenance. The various types of facade access systems are shown in Fig. 5.7.

5.5.1. Permanent Systems

A permanent access system is a planned installation which is implemented at the construction stage. These access systems are usually used for maintenance of facades, window cleaning and building inspection for long-term purposes [32–35]. They can be classified as shown in Table 5.9.

Permanently installed gondolas are dedicated to a specific building or structure. These systems are provided to facilitate regular inspection, maintenance and cleaning purposes. As it is specifically designed for each particular building, it introduces an element of safety in external access engineering which is unsurpassed. For high-rise buildings, gondola systems are more frequently used. This system is usually used due to its low installation cost. Nevertheless, this type of system does involve safety disadvantages. It is dangerous when cleaning is done under strong wind conditions.

A permanently installed gondola consists of a platform suspended from a suspension rig. The suspension rig is usually a trolley unit with a

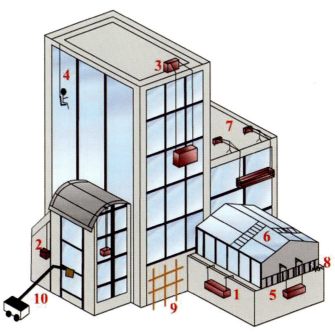

Permanent Access system	Temporary access system
1: Trolley unit/ powered travelling davits	7: Counterweight suspended beam
2: Mono rail system	8: Parapet clamps
3: Traversing trolley	9: Scaffolding
4: Personal units	10: Booms
5: Fixed davits	
6: Travelling ladders	

Figure 5.7. The various types of access systems, both permanent and temporary.

hoist, operating either on rails or on a suitable surface such as a concrete track.

5.5.1.1. Trolley Unit

The trolley unit is also known as the Building Maintenance Unit (BMU), or a roof-car. It is a machine equipped with an integrated lifting hoist, a jib and a cradle and running on the roof of a building (Fig. 5.8).

Table 5.9. Classification of permanent access systems.

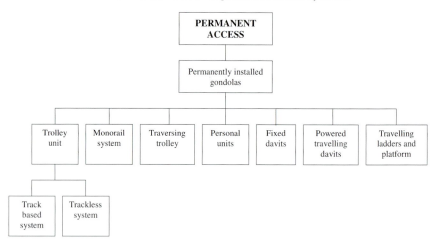

Figure 5.8. A trolley unit suspended from a lifting hoist running on the roof.

It allows complete access to all external parts of the building at the touch of a button. Its movements are controlled safely from the push-button control panel. Access to the cradle and to its garage is from the roof and the operator does not have to assemble or dismantle anything. The built-in safety devices control and monitor the operations of the unit to ensure safety.

The most commonly used trolley unit system used in the region are:
- Track based system (Fig. 5.9).
- Trackless system (Fig. 5.10).

5.5.1.2. Monorail Track

This system is used mainly for recessed or overhanging facades, buildings with sloped roofs and for cleaning the inside of a glazed atrium.

The monorail track follows the line of the facade closely, with cradles suspended from manual or powered trolleys to reach the various points of the facade or roof. It also permits access to inclined roof sections (Figs. 5.11 and 5.12).

(a)

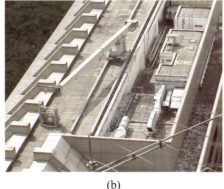

(b)

Figure 5.9. Trolley unit system running on tracks.

(a)   (b)

Figure 5.10. Trackless trolley unit running on wheels.

Figure 5.11. Monorail track along the line of the facade.

5.5.1.3. Traversing Trolley

Depending on the design of the building or the architect's requirement, a traversing trolley may be considered for horizontal traversing on unreachable areas (Fig. 5.13).

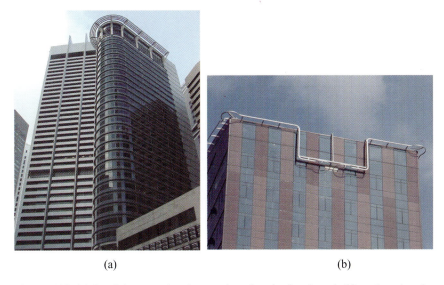

(a)   (b)

Figure 5.12. (a) Special support brackets anchored to the facade and; (b) anchored to the roof.

Figure 5.13. Traversing trolley along glazed.

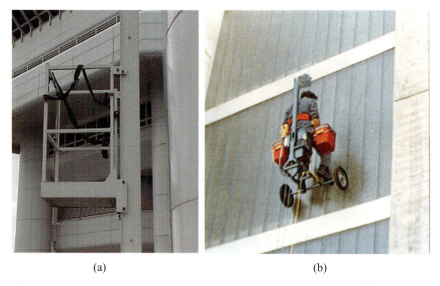

(a)   (b)

Figure 5.14. (a) A personal unit anchored on a ladder for vertical cleaning and (b) A one-man working seat.

5.5.1.4. Personal Unit

Personal units are access systems that allow access to the facade for only one man. They are usually used when the area to be cleaned is relatively small, or for minor cleaning and repair jobs such as window cleaning (Fig. 5.14).

5.5.1.5. Fixed Davit

Fixed davits can be mounted on the roof slab or parapet wall. The manner of mounting will determine the ease by which the cradle can be moved from one section to another to reach the different areas of the building's facade. The davits can be rotated 180° to launch it off the roof onto the facade (Fig. 5.15).

5.5.1.6. Powered Travelling Davit

The difference between the powered travelling davit and the fixed davit

(a)   (b)

Figure 5.15. Platform suspended from fixed davits.

is that the travelling davit is mounted on rails fixed to the parapet (Fig. 5.16). This allows the davit to move from one working position to another easily. This system is especially useful when the roof is particularly crowded with services.

5.5.1.7. Travelling Ladders and Platforms

These are vertically mounted or sloped, guided along rails and designed to blend in with the shape and colour of the background of the building (Figure 5.17 and 5.18). It is particularly useful for facades or building envelopes that are odd in shape and not able to take loadings.

5.5.2. Temporary Access

Temporary access systems can be classified as follows (Table 5.10) [32–35]:

5.5.2.1. Temporary Installed Gondolas

Temporary installed gondolas are temporarily fixed on a building or structure for carrying out a specific task. It consists of a platform and a suspension rig which are assembled prior to use on a work site. They are dismantled and removed from the site on completion of the work.

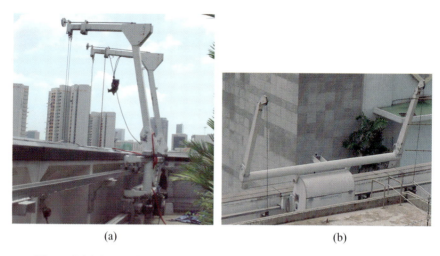

(a)   (b)

Figure 5.16. Powered travelling davit running on rails fixed on parapet wall.

Figure 5.17. Traversing ladder with fixed guardrails for cleaning the exterior of a glazed roof.

Figure 5.18. Traversing ladder for cleaning the inside of a glazed roof.

Table 5.10. Classification of temporary access systems.

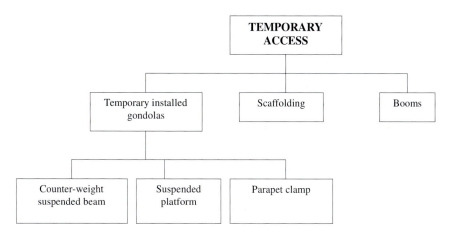

It is usually used by operators for external wall repairs, painting, maintenance and inspection works.

5.5.2.2. Counterweighted Suspension Beam

It consists of counterweight to stabilise the gondolas. The loading

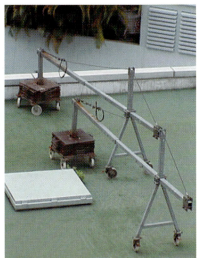

Figure 5.19. Mobile suspension jibs with counterweight.

allowed on the gondola will thus be determined by the counterweights used (Fig. 5.19).

5.5.2.3. Suspended Platform

Suspended platforms are platforms that are fixed temporarily on davits (jibs). They are normally used for maintenance work in residential building or new building constructions and can be moved from one place to the other. The platforms used can also be double-decked for increased efficiency and simultaneous work to different storey heights (Fig. 5.20).

5.5.2.4. Parapet Clamp

If a sufficiently solid and strong parapet (reinforced concrete or steel) is available, the parapet clamp may be used. The stability of the clamp is provided by the parapet itself. The parapet will thus need to be certified to take the loading required (Fig. 5.21).

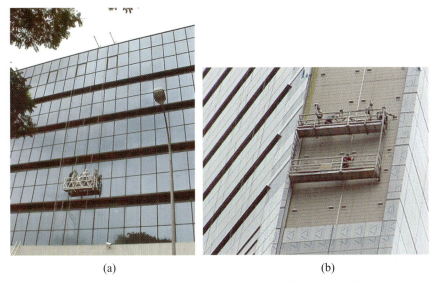

Figure 5.20. Temporary suspended platform on davits. (a) Single deck. (b) Double deck.

Figure 5.21. Parapet clamp as used on a parapet wall.

5.5.2.5. *Scaffolding*

It is mobile and can be temporarily assembled along the perimeter of the buildings. Scaffolding has the advantage of being light-weight, easy, quick to erect and economical (Figs. 5.22 and 5.23). For these reasons, they are widely used in the region for new construction as well as for

(a)   (b)

Figure 5.22. (a) Mobile scaffolding. (b) Overhanging scaffolding.

Figure 5.23. Bintagor scaffolding is very popular among developing countries due to its low cost and sturdy framing members.

refurbishment works. It is seldom used for cleaning and maintenance works due to its limited reach and its difficulty in traversing.

5.5.2.6. Boomlifts

For low rise cleaning and maintenance works of up to 60 metres, boomlifts are most often used. They are extremely mobile and can be hired on contract basis. A strong foundation and unobstructed access path is required when using the boomlift (Fig. 5.24).

Figure 5.24. Articulated boomlifts for external facade's maintenance.

5.5.2.7. Facade-cleaning Robots

The use of robotics technology can be used to overcome the dangerous and time consuming nature of cleaning work, especially that of facade cleaning (Fig. 5.25). Cleaning systems can be specially designed and built to tailor to the specific requirements, such as the surface and geometry of a facade [36, 37].

The advantages of utilising cleaning robots for facades include:

- Complete system for automatic facade cleaning.
- Low operating costs with no personnel costs.
- Constant availability that enhances flexible cleaning cycles, cleaning of especially dirty facade areas or cleaning on demand.
- Usable on various types of facades.
- Secure movement without guide rails on the facade.
- Easy to operate.
- Efficient, economical and ecological cleaning if recycled water is used.
- Short cleaning times for large surfaces.

Figures 5.25. Cleaning robots on glass facades (Courtesy of: Fraunhofer Institute for Factory Operation and Automation IFF).

5.6 Building Forms

A building's overall form has a direct impact on the ease of cleaning of building facades. Buildings with simple and conventional tower forms are likely to incur lower cleaning costs since accessibility to the facade would be easier. On the other hand, access system of buildings with amorphous shapes may not allow workers full and safe access to the facade. It is common to find buildings whose access systems do not offer adequate coverage to its external wall areas but yet are expensive and dangerous to maintain and operate. Such systems are usually designed at later stages of the project where they are made to fit the form of the building.

In ensuring maximum and efficient coverage of external wall areas, the conceptualising of a building's overall form should be done with adequate consideration to the design, provision and future operation of its access systems [30–33]. Table 5.11 illustrates common building forms and their impacts on the provision of access systems [38].

Table 5.11. Common building forms that may affect the provision of access systems.

Building Form	Impacts on Provision of Access System
Plain rectangular 	• Regular and plain form allows ease in the provision of access system. • Roof services may be located in the centre of the roof, allowing tracks or rails for the access system to be laid along the perimeter. • Minimum number of corners to turn. • Maximum coverage with numerous types of access system.
Plain circular 	• Plain form allows ease in the provision of access system. • No corners present. Access system can operate without need to be raised to turn corners. Maximum efficiency. • Working platform should be contoured to the curvature of the building. • Tracks can be laid along the circumference of the roof for the access system to run on.
Sloping face 	• Small roof space due to building form. • The sloping face may be accessed with a jib that extends together with the outward sloping face as the working platform lowers. • The extendable length of the jib should be able to cover the face when the face is at its largest width.
Stepped roof 	• Roof space for services and access system is reduced. • Different access systems may need to be provided on each roof level in order to provide full coverage of facade. • Intermediate roofs should have adequate roof space to operate the access system. • Jibs that are located on the highest roof level and are able to extend to reach all areas of the facade will be efficient and cost effective.

Table 5.11. (*Continued*)

Building Form	Impacts on Provision of Access System
Staggered face 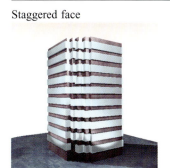	• Numerous corners require the access system to change over at each turn of corner. This may slow down maintenance works and hence reduce efficiency. • Each staggered face should be wide enough to accommodate the access system. • The access system should be equipped with luffing jibs to enable ease when turning the corner.
Sloping roof 	• Minimal roof area for storage and operation of access system. • The track or rail for the access system has to be inclined to the gradient of the roof slope. • If roof is glazed, the access system must be light-weight and within the loading that the roof can withstand.
Pointed roof 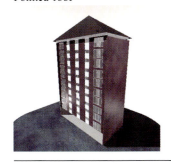	• Minimal roof area for storage and operation of access system. • Access system may be anchored to facade or housed within the roof space that is provided with openings for operating the access system. • Such building forms are usually provided with a flat area at a lower roof level, that runs around the perimeter. • Access to the pointed roof area is possible with a traversing trolley or human rapellers.

References

[1] Smith D. [1988], *"Reliability and maintainability in perspectives"*, Macmillan Education Ltd.

[2] A.C. Lemer and F. Moavenzadeh [1972], *"Performance concept in buildings"*, edited by Secretary, National Bureau of Standards Special Publication 361, Vol 1, Joint symposium committee, Building Research Division, Institute for Applied Technology, NBS, Washington, D.C.20234.

[3] M.Y.L. Chew, Nayanthara De Silva, S.S. Tan (2003). Maintainability of buildings in the tropics, *International Workshop on Management of Durability in the BuildingProcess*, Politecnico di Milano, Italy, 25–26 June 2003.

[4] Bargh J. [1987], *"Increasing maintenance needs on limited budgets - a technique to assist the building manager in decision making"*, in Building Maintenance Economics and management edited by Spedding, E & F.N. Spon publisher.

[5] Building Performance Group [1998], *"Technical audit of building and component durability"*, London.

[6] M.Y.L. Chew, Nayanthara De Silva, P.P. Tan (2003). Maintainability of facades in the tropics, *Conference for Facade Design and Procurement*, Center for Window & Cladding Technology, University of Bath, UK.

[7] Flanagan Roger, Norman George, Meadows Justin, and Robinson Graham [1989], *"Life cycle costing, theory & practice"*, BSP Professional Books, UK.

[8] Housing Association Property Manual (HAPM) [1997], *"Quality scores"*, in Technical Notes April 1999 issue, London.

[9] John R. Meier and Jeffrey S. Russell [2000], "Model process for implementing maintainability", *Journal of Construction Engineering and Management*, November 2000.

[10] Knezevic J. [1993], *"Reliability, maintainability and supportability: a probabilistic approach"*, McGraw-Hill.

[11] Venning Cyril [1995], *"External access systems"*, in Building Maintenance Technology in Tropical Climates edited by Cliff Briffet, Singapore University Press.

[12] Singapore Productivity and Standards Board (1997), Singapore Standard "CP 67 Code of practice for Cleaning and Surface Repair of Buildings, Part 1: Natural Stone, Cast stone and Clay Brick Masonry", Singapore.

[13] Grimm, C.T., *"Cleaning Masonry — A Review of the Literature"*, Construction Research Center, University of Texas at Arlington, 1988.

[14] Mack, R.C., *"The Cleaning and Waterproof Coatings of Masonry Buildings"*, *Preservation Briefs No. 1*, National Park Service, Washington, D.C., 1975.

[15] M. Marosszeky and M.Y.L. Chew, "Importance of workmanship on concrete durability", Concrete *Durability seminar, National Building Technology Center, Sydney, September 1987.*

[16] Singapore Productivity and Standards Board (1999), Singapore Standard "CP 67 Code of practice for Cleaning and Surface Repair of Buildings, Part 2: Concrete and concrete masonry", Singapore.

[17] "Cleaning Concrete Surfaces", REMR Technical Notes CS-MR-4.4, United States Army Engineer Research and Development Center, Vicksburg, MS.

[18] A.M. Sowden, *Maintenance of brick & stone masonry structures*, E & F.N. Spon, London, 1990.

[19] M.Y.L. Chew, L.H. Kang, C.W. Wong, *"Building Facades: A Guide to Common Defects in Tropical Climates"*, World Scientific, 1998.

[20] D. Ramsey, Tile floors, Blue Ridge Summit, PA, Tab Books, 1991.

[21] B.B.P. Lim, Control of the external environment of buildings: selected papers on the protection of the external surfaces of buildings in warm humid climate, Singapore University Press, National University of Singapore , 1988.

[22] "Maintenance of Anodic and powder coating on architectural aluminium", Association of Architectural Aluminium Manufacturers of South Africa, 2000.

[23] C.J. Thomas, *Twentieth-century building materials: history and conservation*, McGraw-Hill, New York, 1995.

[24] L.G.W. Verhoef, *Soiling and Cleaning of Building Facades*, Report of the Technical Committed 62 SCF, RILEM, Chapman and Hall, London, 1988.

[25] A. Field, *Home Maintenance and repair,* Michigan State University Extension, 1998.

[26] J.S. Armstock, *Handbook of Glass in Construction*, McGraw-Hill, New York, 1997.

[27] Building and Construction Authority [2001], Good Practice Guide for Paint.

[28] D. Greay, *Interior and Exterior Painting.* Reston Publishing Company, Va, 1979.

[29] J.H. Arnison, *Floor and Structural Surfaces,* Butterworths, London, 1969.

[30] C. Briffett, Design feedback — successes and failures. In A. Spedding (Ed.), *Building maintenance economics and management* (pp. 215–233). London: E & F.N. Spon, 1987.

[31] E.G. Lovejoy, Safety and security in accessibility for maintenance. In E.D. Mills (Ed.), *Building maintenance and preservation* (pp. 117–130). London: Butterworths, 1980.

[32] J.D. Harrison, Access requirements for building maintenance. In C. Briffett (Ed.), *Building maintenance technology in tropical climates* (pp. 117–138). Singapore: Singapore University Press, 1995.

[33] D. Miles and P. Syagga, *Building maintenance* (pp. 3–7). Intermediate Technology Publications, 1987.

[34] G.T. Hall, Revision notes on building maintenance and adaptation. London: Butterworths, 1984.

[35] I.H. Seeley, *Building maintenance* (20–23). London: Macmillan, 1990.

[36] A. Warszawski, Industrialization and robotics in building — A managerial approach. Harper & Row, 1990.

[37] J. Bohme, N. Elkmann, T. Felsch, J. Hortig, M. Sack, and J. Saenz, Modular climbing robot for service-sector applications. *Industrial Robot: An International Journal*, 26(6), 460–465, 1999.

[38] Council on Tall Buildings and Urban Habitat, *Tall building systems and concepts*. New York : American Society of Civil Engineers, 1980.

INDEX